Adsorption

Progress in Fundamental and Application Research

Adsorption

Progress in Fundamental and Application Research

Selected Reports at the 4th Pacific Basin Conference on Adsorption Science and Technology

Tianjin, China 22 – 26 May 2006

editor

Li Zhou

Tianjin University, China

World Scientific

NEW JERSEY • LONDON • SINGAPORE • BEIJING • SHANGHAI • HONG KONG • TAIPEI • CHENNAI

Published by

World Scientific Publishing Co. Pte. Ltd.
5 Toh Tuck Link, Singapore 596224
USA office: 27 Warren Street, Suite 401-402, Hackensack, NJ 07601
UK office: 57 Shelton Street, Covent Garden, London WC2H 9HE

British Library Cataloguing-in-Publication Data
A catalogue record for this book is available from the British Library.

ADSORPTION
Progress in Fundamental and Application Research

ISBN-13 978-981-277-025-7
ISBN-10 981-277-025-9

Printed in Singapore by World Scientific Printers (S) Pte Ltd

FOREWORD

Adsorption-based technology has experienced a considerable change during the past 30 years from a relatively minor technique to a major one that industry, such as chemical or petrochemical, gaseous or liquid separation and/or purification, relies on today following the progress achieved in the fundamental research, development of novel adsorbents, new adsorption processes, and in combination with other processes, which implies a great potential of decreasing industrial cost. The present book, composed of selected papers of the 4th Pacific Basin Conference on Adsorption Science and Technology held in Tianjin, China for May 22-25, 2006, reflects partially the present state of the art.

Taking on the conference opportunity, about a hundred researchers got together from 18 countries or districts to exchange the recent achievements in adsorption research. However, a conference is indeed an information fair, whose function is more informative than educative. In addition, some papers might not be well organized/written due to the language problem. Therefore, instead of a full proceeding, a collection of contributions is published in the monograph. It is pitiful that some well known scholars could somehow not come to the conference, yet quite a few authors of the monograph are well known for the world adsorption community due to their publication and contribution to the progress of adsorption in the past years. Therefore, what presented in this monograph may attract the attention of adsorption researchers and do benefit their job. It is also desired that some points of view put forward in the book will consequence in more discussion or disputation, as such, real contribution is made to the future development.

Li Zhou

Organizer of the 4-PBAST
Professor and director of
High Pressure Adsorption Laboratory
School of Chemical Engineering and Technology
Tianjin University, Tianjin, China
E-mail: zhouli@tju.edu.cn; zhouli-tju@eyou.com
www.hpal-tju.com

CONTENTS

Part A: General

ADSORPTION KINETICS: THEORY, APPLICATIONS AND RECENT PROGRESS

DOUGLAS M. RUTHVEN

Department of Chemical and Biological Engineering University of Maine, Orono, ME, 04469, USA

E-mail druthven@umche.maine.edu

Over the past thirty years adsorption separation technology has developed from a relatively minor niche process to a major unit operation, with adsorption processes in widespread use in the petroleum and petrochemical industries and in the production of industrial gases as well as in more traditional applications such as air and water purification. The impact of improved understanding of the interplay between adsorption, micropore diffusion and reaction on the development of zeolite catalyzed processes has been even more dramatic. These developments have been stimulated by a dramatic increase in adsorption research which has led to major discoveries ranging from new microporous adsorbent materials to new theoretical approaches yielding improved understanding of adsorption and diffusion in porous solids. Since a comprehensive review is not possible in a single lecture this review has been restricted to a limited number of areas in which recent research has led to the development of new processes or to new concepts where future commercialization appears probable.

1. Zeolite Membranes

The possibility of producing thin coherent defect free zeolite membranes that will allow industrially important molecular sieving separations to be carried out as a continuous flow process has attracted much attention over the past decade

Table 1. Zeolite Membrane Separations

System	Membrane Material	Selectivity	Flux (kg/m^2hr)	Ref
H_2O/Ethanol	NaA	$>10^3$	5 - 15	Morigami *et al* [3] Kondo *et al* [4]
Ethanol/H_2O	Silicalite	25	10	Motuzas [5]
CO_2/CH_4	SAPO-34	50	2.5	Li [6,7]
	DDR	200	1.3	Tomita [8]
CO_2/N_2	SAPO-34	16	0.6	Poshusta [9]
C_6H_6/C_6H_{12}	NaX/NaY	100	0.1	Jeong [10]
Px/Mx	Oriented MFI	200	0.05	Lai [11] Hedlund *et al* [12]

[1,2]. Some examples are listed in Table 1. The separation of water from alcohols (and other organics) by pervaporation through a Zeolite A membrane is now commercial and the CO_2/CH_4 separation, which is important for the exploitation of many low grade natural gas wells, appears poised for commercialization.

Permeance and Selectivity

The simplest model for permeation through a zeolite membrane assumes a linear equilibrium isotherm and a constant diffusivity. The driving force is provided by the difference in partial pressure across the membrane so:

$$N = \frac{KD}{\ell}(p_H - p_L) \tag{1}$$

The constant of proportionality between the flux and the pressure difference (KD/ℓ) is commonly referred to as the permeance while the product of the permeance and the membrane thickness (KD) is referred to as the permeability. At low sorbate concentrations (in the linear region of the isotherm) all components of a mixture diffuse independently so the selectivity is given by:

$$S_{AB} = \frac{J_A}{J_B} = \frac{K_A D_A}{K_B D_B} \tag{2}$$

Since the temperature dependences of D and K follow respectively Arrhenius and vant Hoff expressions [$D = D_\infty e^{-E/RT}$; $K = K_\infty e^{-\Delta U/RT}$] the permeance is expected to vary exponentially with reciprocal temperature, either increasing or decreasing depending on the relative magnitudes of E and ΔU. Such behavior is commonly observed at low loadings (see figure 1a) [13]. However at higher loadings the permeance generally passes through a maximum as shown in figure 1b [14].

To understand this behavior it is necessary to recall that the true driving force for diffusive transport is the gradient of chemical potential, rather than the concentration gradient. Assuming an ideal Langmuir isotherm with an ideal vapor phase the flux is given by:

$$N = \frac{D_0 q_s}{\ell} \ell n \left[\frac{1 + bp_H}{1 + bp_L}\right] \tag{3}$$

in place of Eq. 21, where D_0 is the thermodynamically corrected transport diffusivity defined by [15]:

$$D_0 \equiv BRT = D\left(\frac{d\ell nq*}{d\ell np}\right) \tag{4}$$

Eq. 3 correctly predicts that, for given values of the upstream and downstream partial pressures (p_H and p_L) the flux [and therefore the permeance defined as $J/(p_H\text{-}p_L)$] will pass through a maximum with temperature, as commonly observed. Note that at low loadings (bp << 1.0) Eq. 3 reduces to Eq. 1.

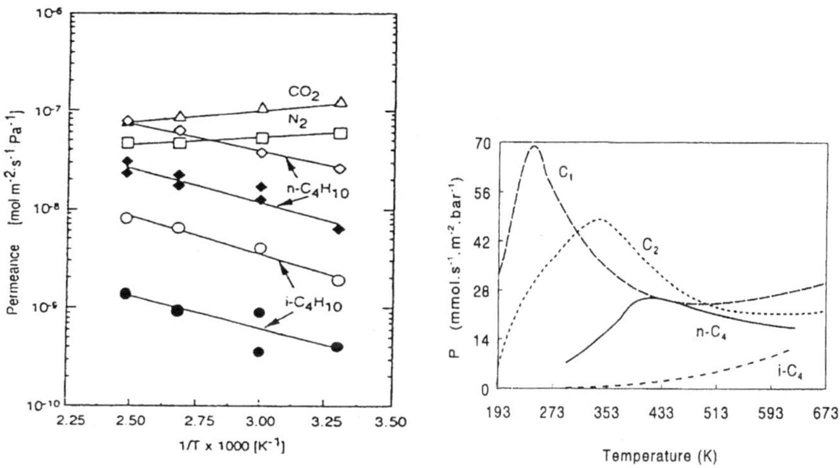

(a) (b)

Figure 1. Temperature dependence of (a) Permeance and (b) Flux for permeation of permanent gases and light hydrocarbons through silicalite membranes.
(a) shows permeance data for N_2, CO_2 and nC_4/iC_4 as a function of reciprocal temperature from data of Kusabe et al [13]. Note that the data for permeation of nC_4 / iC_4 mixtures (filled symbols) show a reduced flux but a higher selectivity suggesting that the permeance of iC_4 is reduced more than that of nC_4 by competitive adsorption.
(b) shows fluxes of CH_4, C_2H_6, C_3H_8 and n/iC_4 plotted as a function of temperature for fixed P_H and P_L taken from data of Bakker *et al* [14].

Permselective Separations

In nanoporous materials diffusion is sterically hindered so that the diffusional activation energy (and hence the permeance) are strongly dependent on molecular size (see Fig. 2), thus giving rise to the possibility of size selective molecular sieve separations. In extreme cases where one of the components is sterically excluded from the pore system a highly efficient molecular sieve separation may be achieved (provided that the membrane is coherent). However, large separation factors are achieved only when the larger molecule is completely excluded. If the larger molecule is small enough to enter the pores, albeit slowly, the perm-selectively drops dramatically since in that situation the conditions for single file diffusion are approached in which all molecules travel at the rate of the slowest. This is illustrated in Table 2 [2].

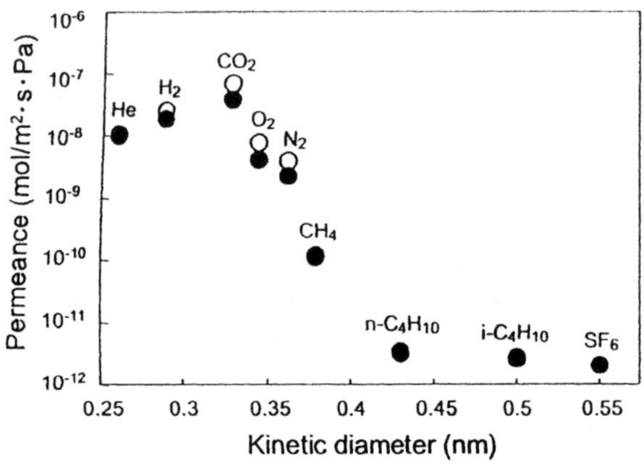

Figure 2. Variation of permeance with kinetic molecular diameter for light gases in DDR type zeolites at 301 K (o) and 373K (•). From Tomita *et al.* [8]

Table 2. Separation pattern of an $AlPO_4$-5-in-nickel-membrane foil at 91°C and 1 bar pressure difference over the membrane. Feed: binary mixtures 1:1 of *n*-heptane and an aromatic compound. (From Caro et al [2]).

	n-heptane (single component)	*n*-heptane/ toluene	*n*-heptane/ mesitylen	*n*-heptane/ triethylbenzene	*n*-heptane/ triisopropylbenzene
Flux x 10^6/mole $s^{-}cm^2$	3.9	0.85	0.43	1.82	0.94
Flux relative to pure *n*-heptane	100%	22%	11%	47%	24%
Selectivity	-	0.8	1.7	105	1220

Interference effects become important only at relatively high loadings so, when there is a large difference in diffusivity between components, both flux and selectivity decrease strongly with loading, as illustrated in Figure 3 [16].

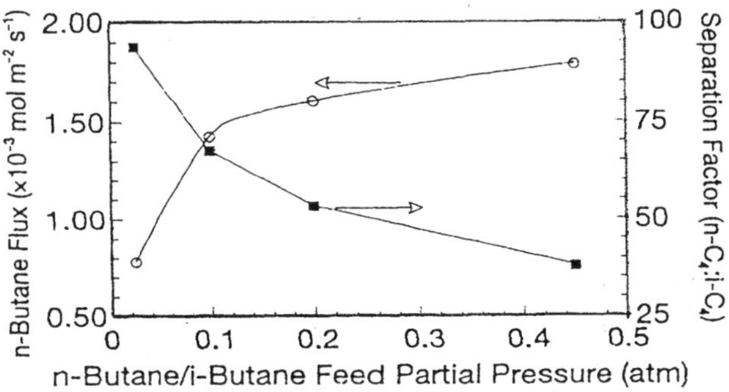

Figure 3. Variation of flux and selectivity with loading for permeation of nC_4 / iC_4 through a silicate membrane. From Tsapatsis *et al* [16].

The perm-selectivity for a mixture is generally found to be lower than the ratio of the pure component permeances (Eq. 2). However, this is not always true. If the faster diffusing species is also the more strongly adsorbed species then, under conditions of competitive adsorption, the adsorption of the slower (and weaker) component will be suppressed by competitive adsorption leading to an *increase* in perm-selectivity [17]. Such an effect has been observed for n-hexane/dimethyl butane in a silicalite membrane for which separation factors in the mixture are greater than 1,000 in favor of n-hexane [17, 18]. This effect is particularly strong for mixtures containing a fast diffusing but weakly adsorbed species (such as H_2) and a more strongly adsorbed but slower diffusing species (e.g. H_2/SF_6 or CH_4/C_4H_{10}) [19, 20].

At high sorbate loadings the effect of differences in adsorption equilibrium tends to become dominant. Thus for methane/n-butane on a silicalite membrane the pure component diffusivity ratio, at ambient temperature, is about three in favor of methane. However, in the binary mixture the selectivity is inverted leading to preferential permeation of n-butane ($S_{CH4/nC4} \approx 0.06$) [21]. The transient behavior of this system is shown in Figure 4. When a clean silicalite membrane is exposed to a 50-50 binary mixture of methane + n-butane the permeate is initially almost pure methane. The butane penetrates the membrane more slowly so that butane appears in the permeate only after about 45 secs. As the butane flux increases the methane flux declines because the strongly adsorbed butane hinders access of the methane to the pores. If the temperature is increased above 200°C the butane loading decreases to a sufficiently low level that methane again becomes the preferentially permeating species.

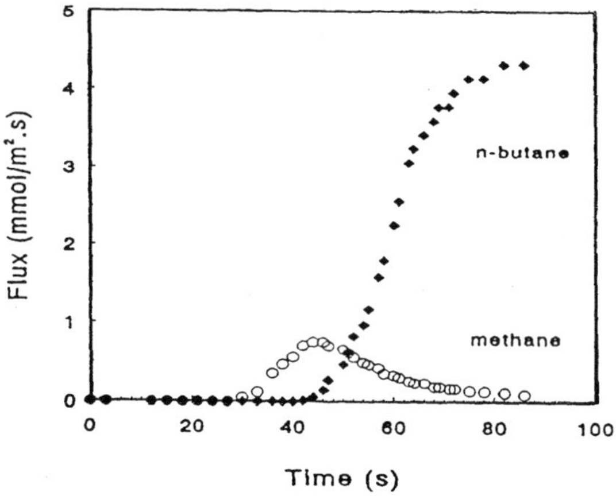

Figure 4. Transient permeation behavior of a 50-50 binary mixture of CH4/nC4H10 in a silicalite membrane at 298K. From Geus et al [21].

Modeling of Permeation in Binary Systems

To properly account for such effects a more sophisticated model is necessary. The most promising approach, developed by Krishna and his associates, is based on the generalized Maxwell-Stefan (GMS) model [22-30]. The basic expression for the flux in a multicomponent system is:

$$-\frac{q_i}{RT}\nabla\mu_i = \sum_{s=i}^{n}\frac{q_j N_i - q_i N_j}{q_s Đ_{ij}} + \frac{N_i}{D_{oi}} \tag{5}$$

where D_{oi} represents the thermodynamically corrected transport diffusivity for component i (defined in accordance with Eq. 7) and $Đ_{ij}$ represents the mutual diffusion co-efficient. For a binary Langmuirian system Eq. 8 reduces to:

$$N_A = \frac{-q_s D_{OA}}{1-\theta_A-\theta_B}\cdot\frac{\left(1-\theta_B+\theta_A D_{OB}/Đ_{AB}\right)\frac{d\theta_A}{dz}+\left[\theta_A+\theta_A D_{OB}/Đ_{AB}\right]\frac{d\theta_B}{dz}}{1+\theta_A D_{OB}/Đ_{AB}+\theta_B D_{OA}/Đ_{AB}} \tag{6}$$

with a similar expression for N_B. When interference between the diffusing species is negligible ($Đ_{AB}\rightarrow\infty$) this reduces to the simplified expression originally derived by Newton, Round and Habgood [31].

The corrected diffusivities (D_{OA}, D_{OB}) can be derived from single component measurements but the mutual diffusivity ($Đ_{AB}$) is not amenable to direct measurement. Krishna has suggested using the Vignes correlation [32] as an estimation method:

$$Đ_{AB} = D_{OA}^{\frac{\theta_A}{\theta_A+\theta_B}} . D_{OB}^{\frac{\theta_B}{\theta_A+\theta_B}} \tag{7}$$

or, for molecules of different sizes the modified form [27]:

$$q_S Đ_{AB} = \left(q_{SB} D_{OA}\right)^{\frac{\theta_A}{\theta_A+\theta_B}} \left(q_{SOA} D_{OB}\right)^{\frac{\theta_B}{\theta_A+\theta_B}} \tag{8}$$

where q_{SA} and q_{SB} represent the saturation capacities for the two components. This development is based on the ideal Langmuir model for adsorption equilibrium. However the theory can be adapted to incorporate any thermodynamically consistent model for the equilibrium isotherm. The development based on the more realistic ideal adsorbed solution theory (IAS) has been presented by Kapteijn *et al* [27].

Representative comparisons between the experimental permeance and selectivity (for CH_4/C_2H_6-silicalite) and the predictions of the GMS model based on single component data are shown in Figure 5 [26]. Also shown are the corresponding predictions from the Habgood model in which mutual diffusion effects are ignored. For the slower diffusing species (C_2H_6) the predicted flux is only marginally altered by mutual diffusion but for the faster diffusing species (CH_4) the effect of mutual diffusion is considerable so that selectivity predictions based on the simplified Habgood model are substantially in error.

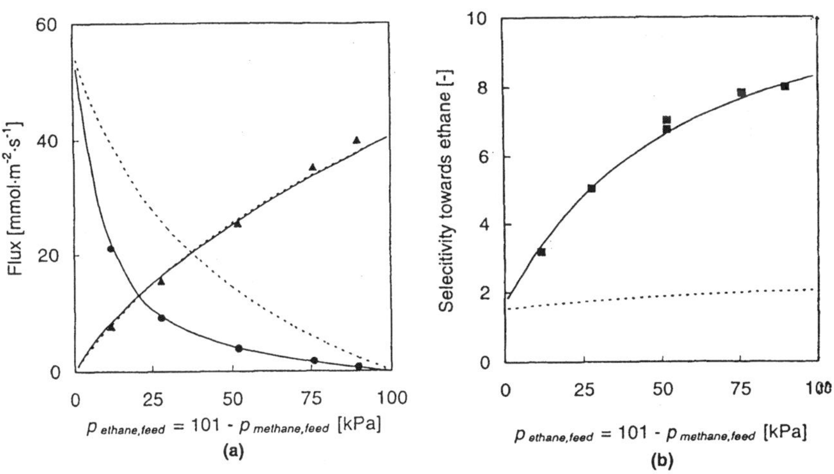

Figure 5. Separation of C_2H_6/CH_4 mixtures by permeation through a silicalite membrane (a) Flux; (b) Selectivity.

Continuous lines show the predictions of the Maxwell-Stefan model (Eq. 9) based on single component values of D_0 with $Đ_{AB}$ estimated from Eq. 11 Dotted lines show predictions of the Habgood model in which mutual diffusion is ignored ($Đ_{AB} \rightarrow \infty$). From van de Graaf *et al* [26].

A similar situation is observed for the separation of CO_2/CH_4 on a SAPO-34 membrane [6,7] (i.e. mutual diffusion leads to higher separation)factors than those predicted from the simplified Habgood model.

A detailed analysis of the influence of mutual diffusion has been carried out by Karimi and Farooq [33]. They show that the effect is generally small at low loadings but becomes important at high loadings when the difference in the mobilities of the two components is large.

Commercialization

Despite their exciting potential the commercialization of zeolite membranes has, so far, been limited. The main barrier appears to be the difficulty of producing sufficiently robust and durable membrane modules of the size required for commercial operation.

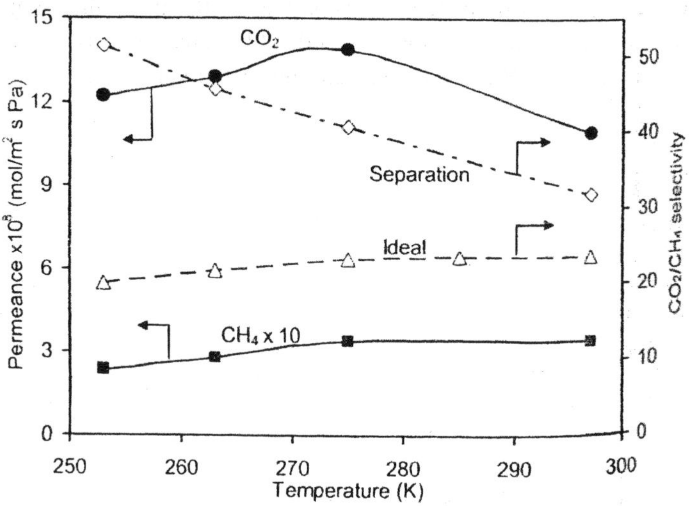

Figure 6. Permeance and selectivity for CO_2/ (50/50 mixture) in a SAPO-34 membrane as a function of temperature. Note: the mixture selectivity is greater than the "ideal" selectivity predicted from single component permeances [6].

2. Kinetic Separations

There are a number of cyclic adsorption separation processes in which the selectivity depends on differences in adsorption rate rather than on differences in equilibrium. Three representative examples of such processes are given below.

Olefin/Paraffin Separations

The separation of light olefins (C_2 H_4 and C_3H_6) from the corresponding paraffins (C_2H_6 and C_3H_8) has traditionally been carried out by cryogenic distillation [34]. However the difference in boiling points is small so the process is energy intensive and therefore costly. The possibility of developing a more competitive adsorption separation process has therefore attracted much research. The earliest such processes took advantage of the fact that, on cationic zeolites, olefins are adsorbed more strongly than the corresponding paraffins [36]. However, the equilibrium selectivity is relatively modest (K_A/K_B ~ 10) and not sufficiently high to achieve a high purity olefin product at high recovery. The possibility of developing an efficient kinetic separation has therefore attracted much recent attention [36-38].

Figure 7 shows diffusivity data for the C_2 and C_3 olefins and paraffins in several different 8-ring zeolites. In 5A zeolite diffusion of the C_2 species is not significantly constrained by steric hindrance so the diffusional activation energy is low (~ 1.5 kcal/mole) with little difference in diffusivity between C_2H_4 and C_2H_6. Steric hindrance is substantially greater in 4A zeolite resulting in higher diffusional activation energies and significantly faster diffusion of C_2H_4, which is the slightly smaller molecule. However, in zeolites of the CHA family, the pores of which are controlled by distorted 8-rings, the differences in diffusivity between olefins and paraffins are much greater (3 to 4 orders of magnitude for C_3H_6/C_3H_8 on high Si CHA). Comparative uptake curves for this system are shown in Figure 8.

The window dimensions and hence the diffusivity and the diffusivity ratio are correlated with the unit cell size. Si CHA, which has the smallest cell size, has the highest kinetic selectivity but the diffusion of propylene is rather slow, thus restricting the cycle time. The choice between a high selectivity with slow uptake of propylene and a lower selectivity with faster uptake thus represents an interesting optimization problem.

Air Separation on Carbon Molecular Sieves

Carbon molecular sieves (CMS) adsorbents are produced by pyrolysis of carbonaceous materials followed by carefully controlled deposition of carbon within the pores [43]. In contrast to activated carbons which have a broad distribution of micropore size (generally in the 10 – 100 Å range) the pores of a carbon molecular sieve are very small (< 10 Å) and the pore size distribution in narrow. As a result the adsorption behavior is similar to that of a zeolite. Carbon molecular sieves are widely used for production of nitrogen from air (by selective adsorption of oxygen). There is little difference between the equilibrium isotherms of O_2 and N_2 on CMS but as a result of its slightly smaller molecular size oxygen is adsorbed very much faster (diffusivity ratio 10 – 100). The sorption kinetics show some interesting features.

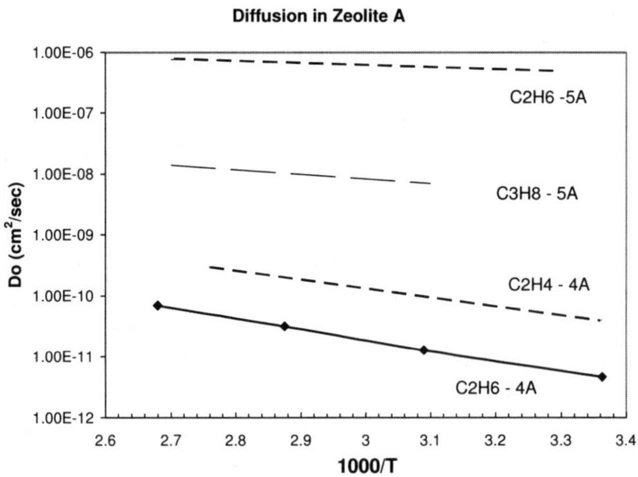

(a)

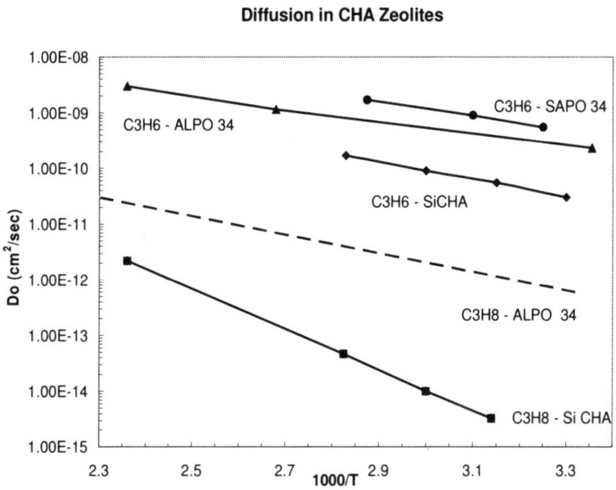

(b)

Figure 7. Arrhenius plot showing the temperature dependence of intracrystalline diffusivity for C_2 and C_3 hydrocarbons in 8-ring zeolites (a) 4A and 5A, (b) CHA zeolites. Data are from refs 36-38 (CHA) and 39-42 (A).

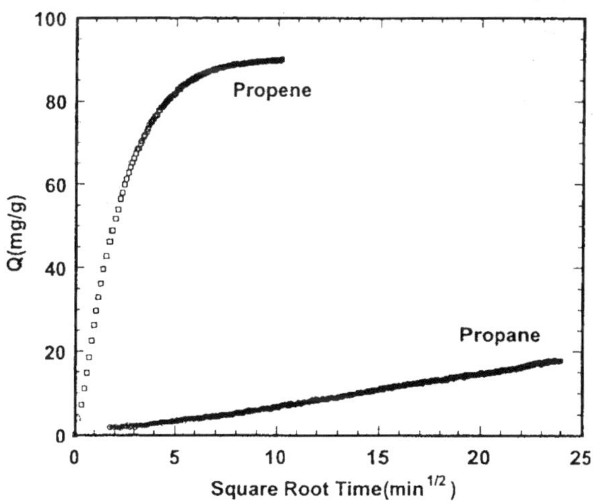

Figure 8. Comparative (integral) uptake curves for C_3H_6 and C_3H_8 in SiCHA at 80° C, 600 Torr. From Olson *et al* [37]. Note that the curves show linearity in $\sqrt{t}$ in the initial region as expected for diffusion control.

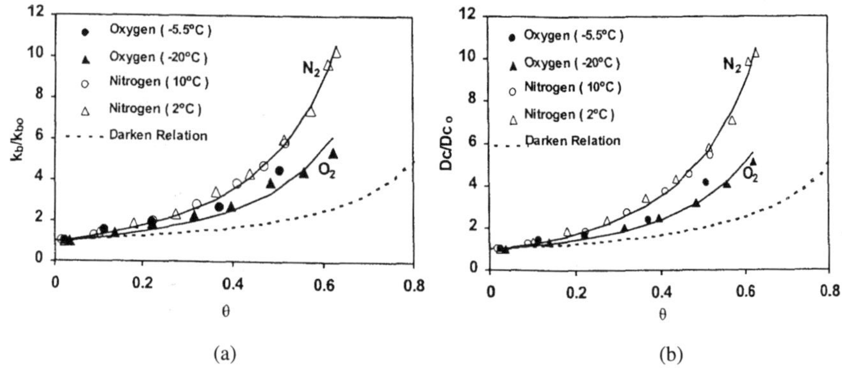

Figure 9. Variation of (a) surface mass transfer coefficient and (b) internal diffusivity with loading for O_2 and N_2 in BF CMS at 298K. From Sundaram *et al*[46].

Detailed studies show that the sorption kinetics are controlled by a combination of surface resistance and internal diffusion although, depending on the particular adsorbent and the conditions, one or other of these resistances may be dominant [44-47]. The uptake curves show a clear transition from surface barrier control in the initial region to diffusion control at long times. The differential diffusivity and the surface mass transfer coefficient both increase strongly with loading; much more strongly than is predicted by the thermodynamic correction factor (Eq. 4). The data are correlated by the empirical expressions:

$$\frac{D}{D_0} = 1 + \beta \frac{\theta}{1-\theta}; \quad \frac{k}{k_0} = 1 + \beta^1 \frac{\theta}{1-\theta} \tag{9}$$

where for N_2 $\beta = \beta^1 = 1.8$ and for O_2 $\beta = 0.76, \beta^1 = 0.89$. Note that for $\beta = 0$ these expressions reduce to the Darken correction for a Langmuir isotherm since $d\ell nq/d\ell np = 1\text{-}\theta$ (see Eq. 4). The physical explanation of this behavior has not yet been established.

N_2/CH_4 Separation over ETS-4

Titanosilicalites such as ETS-4 represent a new class of crystalline microporous molecular sieves, similar to zeolites in their general structure but significantly different in their composition. Like the small pore zeolites ETS-4 has a three dimensional channel structure controlled by 8-membered oxygen rings but the dimensions of the unit cell and hence both the size and shape of the 8-ring windows change dramatically with the dehydration temperature [48]. Provided that the thermal stability limit (~ 200°C for Na form, 330°C for Sr form) is not exceeded this effect is reversible. This flexibility endows these adsorbents with a unique "tuneability" that allows the dimensions of the molecular sieve to be optimized to achieve a particular separation (see Fig. 10). So far the most important industrial application of these materials is in the purification of nitrogen rich natural gas (CH_4).

To meet the calorific value specification for pipeline grade gas the nitrogen content must not exceed about 4%. Many deposits of natural gas, however, contain much larger concentrations of nitrogen. Cryogenic distillation is uneconomic and on both zeolite and CMS adsorbents N_2 and CH_4 are similarly adsorbed with respect to both equilibrium and kinetics, so the search for an economically viable process for nitrogen removal presented the gas industry with an important challenge. The use of ETS-4 dehydrated at 270°C, appears to be a promising solution since this material shows a high kinetic selectivity for N_2 over CH_4 (see Figure 11), thus allowing an effective kinetic separation to be achieved [50]. Following successful pilot plant trials a full scale unit has been developed using a relatively fast cycle (time scale of minutes) pressure swing adsorption process. About 75% of the N_2 is removed with 95% recovery of CH_4. However, the process is not without its problems:

1. The capacity of the adsorbent is relatively low so a large volume of adsorbent is needed.
2. It is essential to dry the feed gas to very low humidity levels.
3. Methane diffuses into the structure albeit slowly, necessitating periodic thermal regeneration of the adsorber beds. This adds significantly to the process cost.

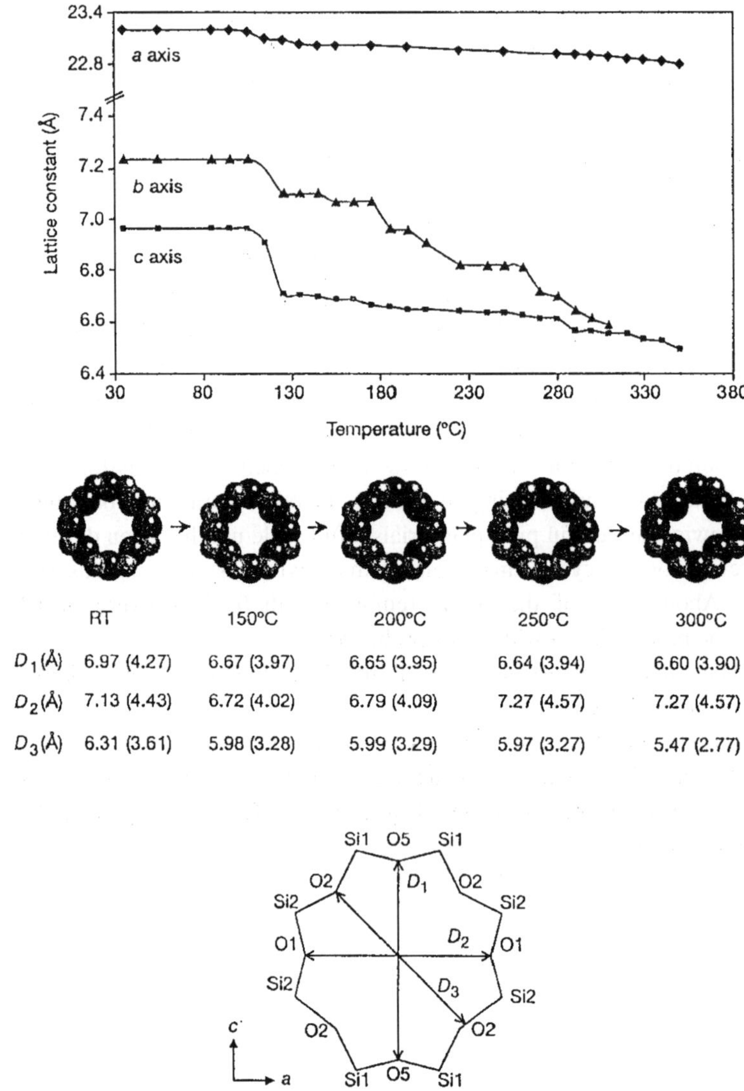

Figure 10. Variation of lattice parameters and pore dimensions of ETS-4(Sr) with dehydration temperature. Modified from Kuznicki *et al*[48].

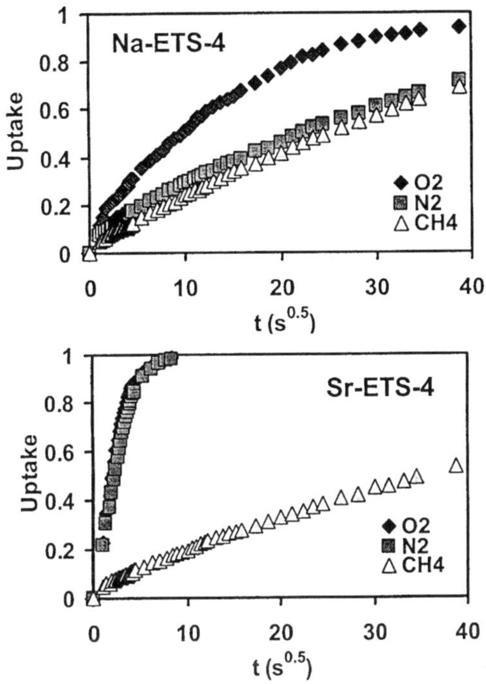

Figure 11. Uptake curves for O_2, N_2 and CH_4 in SrETS-4 (dehydrated at 270°C). Data from Farooq *et al*[49].

3. Diffusion and Catalysis

Catalytic Effectiveness Factors

Diffusion plays a major role in influencing both the activity and selectivity of many catalysts. For a first order reaction in a spherical catalyst particle the intrinsic rate constant (k) is reduced by a factor η (the effectiveness factor):

$$k_e = k\eta$$

$$\eta = \frac{3}{\Phi}\left[\frac{1}{Tanh\Phi} - \frac{1}{\Phi}\right] \tag{10}$$

$$\Phi = R\sqrt{k/D}$$

This basic analysis is commonly attributed to Thiele (1938) [51] and the dimensionless parameter Φ is commonly called the Thiele modulus although essentially the same analysis was published many years earlier by Jüttner [52].

In a zeolite catalyst diffusional limitations may occur at either the particle scale or the crystal scale. In the latter case the basic analysis remains the same but since the rate constant is defined with respect to the concentration of reactant in the vapor phase while the intracrystalline diffusivity is defined with respect to the adsorbed phase concentration, the Thiele modulus must be re-defined to introduce the dimensionless adsorption equilibrium constant (K):

$$\Phi_s = R\sqrt{k/KD} = \left(\frac{R^2}{D}\cdot\frac{k}{K}\right)^{1/2} \quad (11)$$

Both the intrinsic rate constant and the effective diffusivity (KD) can be extracted from measurements of the reaction rate with different size fractions of the zeolite crystals. This approach has been demonstrated by Haag [53] for cracking of n-hexane on HZSM5 and by Post *et al* [54] for isomerization of 2,2 dimethyl butane over HZSM-5.

Catalytic Cracking

Kortunov *et al* [55] have used the PFG NMR technique to measure the diffusion of linear alkanes within the crystals *and* within the macropores of HY and REY based cracking catalysts. At 600°C $D_{macro}/D_{micro} \sim 10$ but, since the crystal size is about 1 μm while the particle size is about 100 μm the ratio of the diffusional time constants $[(r^2/D_{micro})/(R^2/D_{macro})]$ is of order 10^{-3}, showing that under reactor conditions the mass transfer rate is controlled by intraparticle diffusion rather than by intracrystalline diffusion. As a result the performance of a series of industrial cracking catalysts correlates closely with the effective macropore diffusivity. Stallmach and Crowe [56] have shown how the effective macropore diffusivity at certain temperatures may be predicted from PFG NMR measurements at lower temperatures under non-reacting conditions. Their technique provides an *in situ* measurement of the tortuosity factor for the macropores as well as the distribution of sorbate between the zeolite crystals and the macropores.

MTO Reaction

The methanol to olefins (MTO) reaction offers an important example of a catalytic reaction controlled by intracrystalline diffusion. Stimulated by the escalating demand for light olefins, this reaction has attracted much recent attention. The reaction of methanol and 350-450°C over HZSM5 yields a wide spectrum of products including light alkanes, light olefins and single ring aromatics [57-59]. The yield of $C_2^= + C_3^=$ (the desirable product for polyolefin feedstock) amounts to only 30 – 40 %. The introduction of SPO-34 (a structural analog of chabazite) as the catalyst [60] gave a dramatic improvement in both selectivity and conversion, making the process much more attractive. Under properly selected conditions light olefin yields ($C_2^= + C_3^=$) approaching 80% can be achieved with only small amounts of higher olefins and paraffins and essentially no aromatics [61].

The absence of aromatic products appears to be related to the size of the CHA cage which is too small to allow the formation of a benzene ring. The reaction mechanism has been established in broad outline [62, 63] although many important details are still not fully understood:

$$\begin{aligned}&1.\quad 2CH_3OH \rightarrow CH_3.OH.CH_3 + H_2O\\&2.\quad CH_3.O.CH_3 \rightarrow C_2H_4 + H_2O\\&3.\quad 1.5\ C_2H_4 = C_3H_6\end{aligned} \qquad (12)$$

Slow polymerization to higher molecular weight species (coke) also occurs. Reaction 3 is reversible and exothermic; this probably accounts for the observed increase in $C_2^= + C_3^=$ yield with temperature.

Detailed studies of the kinetics of this reaction over different size fractions of SAPO-34 crystals together with measurements of the sorption rate and the equilibrium isotherm have been reported by Chen *et al* [64-68]. These data are

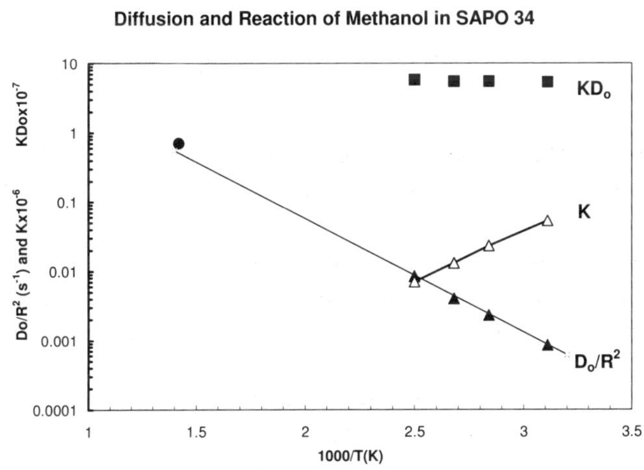

Figure 12. Variation of diffusional time constant (D_0/R^2), dimensionless Henry constant (K) and the product KD_0 with temperature. (From data of Chen *et al* [64]). The value of D_0/R^2 derived from the reaction rate measurements (•) is also shown. Corrected diffusivities are derived from the reported integral diffusivities according to the analysis of Garg and Ruthven [69].

summarized in figure 12. The dominance of intracrystalline diffusion in controlling the sorption rate was shown by varying the crystal size. Values of the diffusional time constant (R^2/D_o) derived from reaction rate measurements at 698K are close to the value extrapolated from sorption rate measurements at lower temperatures with the same batch of SAPO-34 crystals [64, 65]. The temperature dependence of the dimensionless Henry constant, also shown in figure 12, yields an adsorption energy of $\Delta U \approx -7.5$ kcal/mole which is almost the same as the diffusional activation energy derived from the temperature dependence of the (corrected) diffusivity (E = 7.3 kcal/mole.) Consequently the product KD_0, referred to by Chen as the "steady state diffusivity" is almost independent of temperature. A similar situation was noted by Garcia and Weisz [70, 71] in their study of the reaction of various aromatics over HZSM-5.

As the catalyst ages, the light olefin yield and the selectivity both increase [64, 66]. This appears to be related to the build up of coke within the intracrystalline pores which reduces both the intrinsic rate constant and the intracrystalline diffusivity [65, 66]. Detailed measurements with different crystal sizes show that with increasing coke levels the diffusivity declines more rapidly than the rate constant so that diffusional limitations become more pronounced as the catalyst ages. A high yield of light olefins requires that the DME formed in the first step of the reaction be retained within the crystal long enough for it to be essentially fully converted by reaction 2. This requires that the ratio of the Thiele moduli should be large:

$$\frac{\Phi_2}{\Phi_1} = \left(\frac{k_2}{k_1} \frac{D_{MeCH}}{D_{DME}} \right)^{\frac{1}{2}} >> 1 \qquad (13)$$

The *ratio* of the Thiele moduli is independent of crystal size, so in accordance with experimental observations [61], varying the crystal size has no effect on the yield.

Since $k_2 < k_1$ a high ratio of D_{MeOH}/D_{DME} is necessary to achieve a high ratio Φ_2/Φ_1 and thus a high olefin yield. As the DME molecule is larger than the methanol molecule it is reasonable to assume that, under sterically restricted conditions, the diffusivity ratio D_{MEOH}/D_{DME} will increase as the effective pore size decreases. The observations that the olefin yield increases as the catalyst cokes and that an improvement in yield is obtained by increasing the Si/Al ratio (which decreases the unit cell size and therefore the effective window size) are consistent with this hypothesis. However varying the Si/Al ratio also changes the strength of the acid sites so such evidence is not entirely conclusive.

4. Fundamental Studies of Diffusion in Zeolites

The preceding sections provide selected examples showing how sorption and diffusion in zeolite crystals can be exploited to yield technologically useful processes. It is therefore appropriate to conclude this review with a short discussion of the remarkable progress that has been achieved in recent experimental studies of diffusion in zeolite crystals.

Table 3. Experimental Methods for Measuring Intracrystalline Diffusion in Zeolites

- Microscopic Methods (Sub-crystal scale)
  - QENS
  - NMR - Relaxation
    - PFG
  - Neutron Spin-Echo
- Mesoscopic Methods (Single crystal scale)
  - Single crystal Permeation
  - FTIR
  - Interference Microscopy
- Macroscopic Methods (Many crystals)
  - Transient
    - Sorption Rate
      - Flow – ZLC/TZLC
      - Batch – DAB
        - Gravimetric
        - Piezometric
        - FTIR
        - Temp. Response
    - Chromatographic
      - Gas Phase
      - Liquid Phase
      - Wall Coated Column
    - Frequency Response
      - Pressure
      - Pressure/Temperature
  - Quasi Steady State
    - Membrane
      - Wicke Kallenbach
      - Single Crystal
      - Zeolite Membrane
    - Catalyst Effectiveness Factor

For several reasons the reliable measurement of micropore-diffusion has proved to be far more difficult than expected. A wide range of different experimental techniques have been applied (see Table 3). We now know that when the diameter of the diffusing molecule is even slightly smaller than the pore diameter, diffusion within an ideal micropore is surprisingly fast and difficult to measure by macroscopic methods since the size of available zeolite crystals is limited. Such fast processes can, however, be measured relatively easily by PFG NMR and QENS. As the molecular diameter of the sorbate approaches (or even exceeds) the minimum diameter of the pore the diffusional activation energy increases and the diffusivity drops by orders of magnitude. Slow transport-diffusion (for example ethane, propane, etc. in CHA or Zeolite A – see Fig. 7) is easily measured macroscopically but inaccessible to microscopic techniques. The range of systems and experimental conditions where reliable measurements can be made by both macroscopic and microscopic methods is therefore quite restricted.

Transient uptake rate measurements are subject to intrusion of heat transfer limitations, especially in batch measurements at low pressures. Membrane permeation, frequency response and ZLC measurements should not be subject to serious heat transfer limitations but, especially in frequency response and ZLC, there is always a danger of intrusion of extracrystalline resistances to mass transfer, although in principle these can be eliminated by reducing the sample size and ensuring that the crystals within the sample are dispersed rather than aggregated together. Recent measurements have however shown that for many systems significant discrepancies between microscopic and macroscopic diffusion measurements remain even when the intrusion of extracrystalline resistances is carefully minimized. Similarly the diffusivities measured by quasi steady state membrane permeation tend to be larger than the values determined by transient macroscopic methods although still substantially smaller than the microscopic values derived from PFG NMR, QENS and molecular dynamic simulation (see Fig. 13) [72, 73].

A major advantage of the recently developed interference microscopy technique [74, 75] is that in addition to allowing a direct measurement of sorption/desorption rates on the single crystal scale it provides, from the form of the transient concentration profiles, direct experimental evidence concerning the nature of the rate controlling resistances to mass transfer. Recent studies by this technique have shown that the influence of structural defects and surface resistance to mass transfer are far more important than has been generally assumed [76-80]. For some systems it appears that sorption rates are controlled by surface resistance while in other cases the profiles suggest a combination of surface and internal diffusional resistance control – see for example Figure 14 [81]. Sometimes portions of the intracrystalline pore volume are completely inaccessible due to barriers associated with the crystal growth planes. In the case of ferrierite it appears that transport occurs entirely through the 8-ring channels while the larger 10-ring channels provide no access, presumably as a

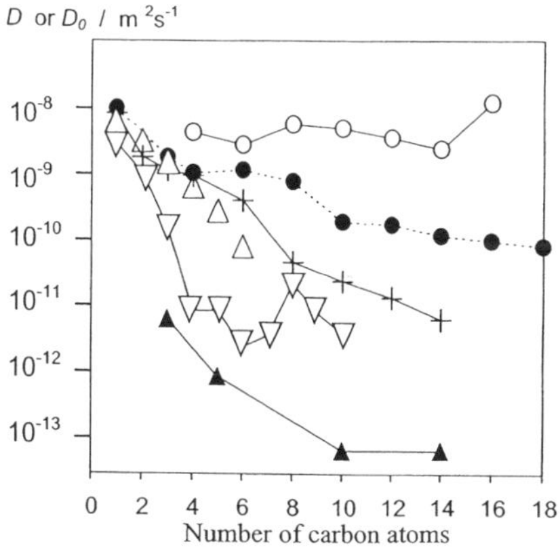

Figure 13. Diffusivities for n-alkanes in silicalite at 300K measured by different techniques. •, o MD simulations; +, QENS; ▽, single crystal membrane; △, PFG NMR; ▲, ZLC. From Jobic [72].

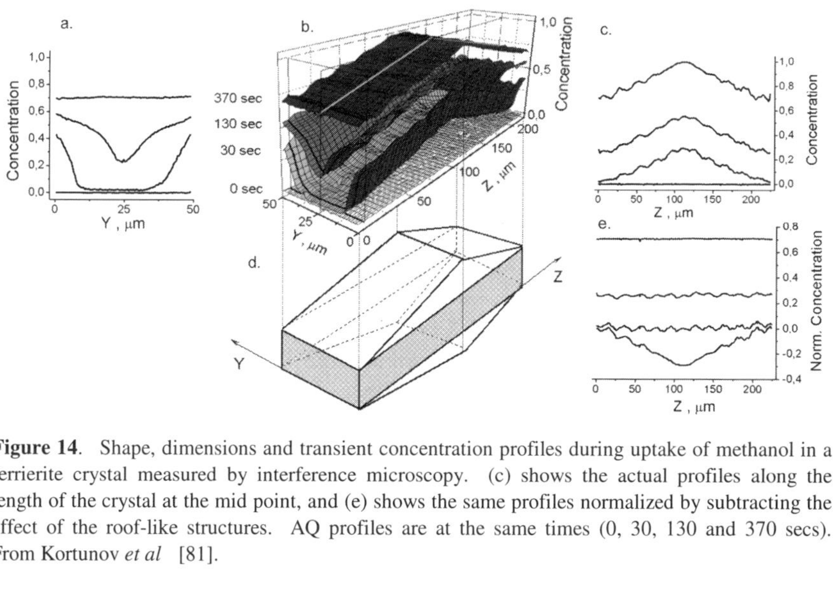

Figure 14. Shape, dimensions and transient concentration profiles during uptake of methanol in a ferrierite crystal measured by interference microscopy. (c) shows the actual profiles along the length of the crystal at the mid point, and (e) shows the same profiles normalized by subtracting the effect of the roof-like structures. AQ profiles are at the same times (0, 30, 130 and 370 secs). From Kortunov *et al* [81].

result of a surface barrier [81]. Less pronounced internal barriers presumably resulting from fault planes within the crystal have also been observed [77].

It thus appears that in real zeolite crystals diffusion over long distances reflects the influence of surface and internal barriers rather than the pore structure of the idealized framework. As a result the apparent intracrystalline diffusivities often show a strong dependence on the length scale of the measurement. Measurements by QENS and neutron spin echo methods over distances corresponding to a few unit cells often approach the theoretical values derived from MD calculations for an ideal lattice. Similar values are often obtained by PFG NMR when the measurement is made over short distances. Measurements by most macroscopic methods are on the length scale of the crystals and these tend to yield lower apparent diffusivities as a consequence of the intrusion of surface barriers and internal resistances due to structural defects.

Measurements by interference microscopy are, under favorable conditions, capable of yielding both internal diffusivities and apparent diffusivities based on overall sorption rates. The former tend to approach the values obtained from microscopic measurements while the latter yield values similar to those obtained by other macroscopic methods. Of necessity these studies have been carried out in large zeolite crystals. One may expect that smaller crystals may be less defective, although the influence of surface resistance may be expected to be greater. The extent to which these conclusions are applicable to the small zeolite crystals generally used in commercial zeolite catalysts and adsorbents remains an important question.

Notation

b Langmuir equilibrium constant (atm^{-1}) q adsorbed phase concentration
B mobility q_s saturation limit
c gas phase concentration of sorbate R particle radius or gas constant
D diffusivity S_{AB} selectivity
D_0 thermodynamically corrected T absolute temperature diffusivity (see Eq. 7)
$Đ_{AB}$ mutual diffusivity
J flux Φ Thiele modulus
k reaction rate constant θ fractional saturation (q/q_s)
K Henry's Law constant β, β^1 constants in Eq. 13
ℓ membrane thickness η effectiveness factor
p partial pressure

References

1. Coronas, J. and Santamaria, J., *Sep. Purif. Methods* **28** (1999) p. 127.
2. Caro, J., Noack, M., Kölsch, P. and Schäfer, R., *Microporous and Mesoporous Mats.* **38** (2000) pp. 3-24.
3. Morigami,Y., Kondo, M., Abe, J., Kita, H. and Okamoto, K., *Sep. and Purif. Technol.* **25**, (2001) pp. 251-260.
4. Kondo, M., Yamamura, T., Yukitake, T., Matsue, Y., Kita, K. and Okamoto, K., *Sep. and Purif. Technol.* **32** (2003) pp. 191-198.
5. Motuzas, J., Julbe, A., Noble, R. D. and Falconer, J – in press
6. Li, S., Falconer, J. L. and Noble, R. D., *J. Membrane Sci.* **241** (2004) pp. 121-135.
7. Li, S., Martinele, J., Falconer, J. L. and Noble, R.D., *I. and E Res.* **44** (2004) p.3220-3228.
8. Tomita, T., Nakayama, K. and Sakai, H., *Microporous and Mesoporous Mats.* **68** (2004) pp. 71-75.
9. Poshusta, J. C., Tuan, V. A., Pope, E. A., Noble, R. D., and Falconer, J. L. *AIChE Jl.* **46**, (2000) pp. 779-789.
10. Jeong, B. H., Hasegawa, Y., Sotowa, K. I., Kusakabe, K. and Morooka, S., *J. Membrane Sci.* **213** (2003) p. 115-124.
11. Lai, Z. P., Bonilla, A. G., Diaz, I., Nery, J. G., Sujaoti, K., Amat, M. A., Kokkoli, E., Terasaki, O., Thompson, R., W., Tsapatsis, M. and Vlachas, D. G., *Science* **300** (2003) p. 455-460.
12. Hedlund, J. H., Sterte, J., Athonis, M., Bons, A. – J., Carstensen, B., Corcoran, E. W., Cox, D. M., Deckman, H. W., De Gijnst, W., de Moor, P. – P., Lai, F., McHenry, J., Mortier, W., Reinoso, J. J. and Peters, J. *Microporous and Mesoporous Materials* **52** (2002) pp. 179-189.
13. Kusabe, K., Murata, A., Kuroda, T. and Morooka, S., *J. Chem. Eng. Japan* **30** (1997) pp. 72-78.
14. Bakker, W. J. W., Kapteijn, F., Poppe, J. and Moulijn, J.A., *J. Membrane Sci.* **117** (1996) pp. 57-78.
15. Ruthven, D. M. Principles of Adsorption and Adsorption Processes (John Wiley, New York, 1984) p. 125.
16. Tsapatsis, M. and Garda, G. R., *MRS Bulletin* **1** (1999) p. 30
17. Falconer, J. In *Int. Workshop on Zeolite Membranes Post Conf. ICIM 98 Proceedings*, (Gifu, Japan, 1998) p. 21.
18. Coronas, J., Noble, R. D. and Falconer, J. L., *Ind. Eng. Chem. Res.* **37** (1998) pp. 166-176.
19. Noble, R. D. and Falconer. J. l., *Catalysis Today* **25** (1995) p. 209.
20. Vroon, Z. A. E. P., Keizer, K., Gilder, M. J., Verweij, H. and Burgraaf, A. J., *J. Membrane Sci.* **113** (1996) p. 293-300.
21. Geus, E. R., Van Bekkum, H., Bakker, W. J. W. and Moulijn, J. A., *Microporous Mats.* **1** (1993) p. 131-147.

22. Krishna, R. and Wesselingh, J. A., *Chem. Eng. Sci.* **52** (1997) pp. 861-911.
23. Keil, F. J., Krishna, R. and Coppens, M. O., *Rev. Chem. Engg* **16** (2000) pp. 71-197.
24. Krishna, R. and Baur, R., *Sep. and Purif. Technol.* **33**, (2003) pp. 213-254.
25. Krishna, R. and van den Broecke, L. J. P., *Chem. Eng. J.* **57** (1995) pp. 155-162.
26. Van de Graaf, J. M., Kapteijn, F. and Moulijn, J. A., *A.I.Ch.E. Jl* **45** (1999) pp. 497-511.
27. Kapteijn, F., Moulijn, J. A. and Krishna, R., *Chem. Eng. Sci.* **55** (2000) pp. 2923-2930.
28. Krishna, R. and Baur, R., *Chem. Eng. J.* **97** (2004) pp. 37-54.
29. Skoulidas, A. I., Sholl, D. S. and Krishna, R., *Langmuir*, (2003) pp.7977-7988
30. Krishna, R. *Chem.Phys.Letters* **326** (2000) pp. 477-484.
31. Round, G. F., Habgood, H. W. and Newton, R., *Sep. Sci.* **1** (1996) pp. 219.
32. Vignes, A., *Ind. Eng. Chem. Fund.* **5** (1966) pp. 189-199.
33. Karimi, I. A. and Farooq, S., *Chem. Eng. Sci.* **55** (2000) pp. 3529-3541.
34. Eldridge, R. B., *Ind. Eng. Chem. Res.* **32** (1993) pp. 2208-2212.
35. Peterson, D. K., Helfferich, F. and Griep, R K. In *Molecular Sieves. Proc. 1st Int. Zeolite Conf.* (Soc. Chem. Ind., London, 1968) pp. 217-229.
36. Olson, D., U.S. Patent 6,488,741 B2 Dec. 3 2002
37. Olson, D., Camblor, M. A., Villaescusa, L. A. and Kühl, G. H., *Microporous and Mesoporous Materials*, **67** (2004) pp. 27-33.
38. Reyes, S. C. *et al*, U.S. Patent 6,730,142 B2 May 4 2004
39. Xu, Z., Eic, M. and Ruthven, D. M. In *Ninth Int. Zeolite Conf., Montreal 1992. Proceedings,* ed. by von Ballmoos, R., Higgins, J. B. and Treacy, M. M. J. (Butterworth, Stoneham, MA, 1993) Vol. 2, pp. 147.
40. Kärger, J. and Ruthven, D. M., *J.Chem.Soc. Faraday Trans. I*, **77** (1981) pp. 1485.
41. Yucel, H. and Ruthven, D. M., *J.Chem.Soc. Faraday Trans. I*, **76** (1980) pp. 60-70.
42. Sheth, A. C. M.Sc. Thesis, Northwestern University, Evanston IL (1969)
43. Jüntgen, H., Knoblauch, K. and Harder, K., *Fuel*, **60** (1981) pp. 817. See also *Adsorption Sci. and Technol.* **158**, pp.269-283. A.E. Rodigues *et al* eds. Kluwer, Dordrecht (1988)
44. Loughlin, K. F., Hassan, M. M., Fatehi, A. I. and Zakur, M., *Gas Sep. Purif.* **7** (1993) pp. 264-273.
45. Shen, D., Bülow, M. and Lemcoff, N., *Adsorption* **9** (2003) pp. 295-302.
46. Sundaram, S. M., Qinglin, H. and Farooq, S. In *Proc. 7th Int. Conf. on Fundamentals of Adsorption,* Nagasaki, May 2001 ed. By Kaneko, K., Kanoli, H. and Hanzawa, Y. (I.K. International, Shinjuku, Japan, 2002) pp. 779-786.

47. Farooq, S. and Bahtia, S. K. – personal communication
48. Kuznicki, S. M., Bell, V. A., Nair, S., Hillhouse, H. W., Jacubinas, R. M., Braunbath, C. – M., Toby, B. H. and Tsapatsis, M., *Nature* **412** (2001) pp. 720-724.
49. Marathe, R. P., Mantri, K., Srinivasan, M. P. and Farooq, S. *I. and E.C. Res.* **43**, (2004) pp. 5281-5290.
50. Mitariten, M., *American Oil and Gas Reporter* Mar. 2001 pp. 103-104.
51. Thiele, E. W., *Ind.Eng.Chem.* **31** (1939) pp. 916.
52. Jüttner, F., *Z. Phys.Chem.* **65** (1909) pp. 595.
53. Haag, W. O., Lago, R. M. and Weisz, P. B., *Disc. Faraday Soc.* **72** (1982) pp. 317.
54. Post, M. F. M., van Amstel, J. and Kouwenhoven, H. W. In *Proc. 6th Int. Zeolite Conf. Reno, Nevada, 1983* ed. by Olson, D. and Bisio, A. (Butterworth, Guildford, UK 1984) pp. 517.
55. Kortunov, P., Vasenkov, S., Kärger, J. *et al.* In *Diffusion Fundamentals* ed. by Kärger, J, Grinberg, F. and Heitjans, P. (Leipzig University Press, Leipzig, Germany, 2005) pp. 548.
56. Stallmach,F. and Crowe, S., *Ibid.* pp.474.
57. Chang, C. D., Chu, C. T. W. and Socha, R. F., *J. Catal* **86** (1984) pp. 289.
58. Chen, N. Y., Garwood, W. E. and Dwyer, F. G. In *Shape Selective Catalysis in Industrial Operations* (Marcel Dekker, New York, 1989) pp. 233-238.
59. Chang, C. D. In Hydrocarbons from Methanol. *Catal. Revs. – Sci.Eng.* **25** (1983) pp. 1.
60. Kaiser, S. W. U.S.Patent 4,499,327 (1985).
61. Wilson, S. and Barger, P., *Microporous and Mesoporous Materials* **29** (1999) pp. 117-126.
62. Froment, G. F., Dehertog, W. J. H. and Marchi, A. J. In *Catalysis.* ed. by Spivey, J. J. (Royal Soc. Chemistry, London, 1992) Vol. 9 Ch. 1.
63. Dahl, I. M. and Kolboe, S., *J. Catalysis* **149** (1994) pp. 458-464 and **161** (1996) pp. 304-309.
64. Chen, D., Rebo, H. P., Moljord, K. and Holman, A., *Ind.Eng.Chem.Res.* **38** (1999) pp. 4241-4249.
65. Chen, D., Rebo, H. P. and Holman, A., *Chem.Eng.Sci.* **54** (1999) pp. 3465-3473.
66. Chen, D., Rebo, H. P., Moljord, K. and Holman, A., *Ind.Eng.Chem.Res.* **36** (1997) pp. 3473-3479.
67. Chen, D., Rebo, H. P., Moljord, K. and Holman, A., *Chem.Eng.Sci.* **51** (1996) pp. 2687-2692.
68. Chen, D., Rebo, H. P., Moljord, K. and Holman, A., *Studies Surf.Sci.Catalysis* **119** (1998) pp. 521.
69. Garg, D. R. and Ruthven, D. M., *Chem.Eng.Sci.* **27** (1972) pp. 417.

70. Garcia, S. F. and Weisz, P. B., *J.Catalysis* **121** (1990) pp. 294-311.
71. Garcia, S. F. and Weisz, P. B., *J.Catalysis* **142** (1993) pp. 691-696.
72. Jobic, H. In *Recent Advances in Gas Separation by Microporous Ceramic Membranes,* ed. by Kanellopoulos, N. K. (Elsevier, Amsterdam, 2000).
73. Kärger, J. and Ruthven, D. M. In *Handbook of Porous Solids,* ed by Schütt, F., Sing K. S. W. and Weitkamp, J. (Wiley-V.C.H., Weinheim, Germany, 2002)Vol.4 pp.2089.
74. Kärger, J. and Schemmert, U. In *Proc. 2nd Pacific Conf. on Adsorption, Brisbane (2000)* ed. by Do, D. D. (World Scientific, Singapore, 2000) p.324.
75. Schemmert, U., Kärger, J. and Weitkamp, J., *Microporous and Mesoporous Mats.* **32** (1999) pp. 101.
76. Geier, O., Vasenkov, S., Lehmann, E., Kärger, J., Schemmert, U., Rakoczy, R. A. And Weitkamp, J., *J. Phys.Chem.* **105** (2001) pp. 10,217.
77. Vasenkov, S. and Kärger, J., *Microporous and Mesoporous Mats.* **55** (2002) pp. 139
78. Wloch, J., *Ibid.* **62** (2003) pp. 81.
79. Kortunov, P., Vasenkov, S., Chmelik, C., Kärger, J., Ruthven, D. M. and Wloch, J., *Chemical Materials* **16** (2004) pp. 3552.
80. Lehmann, E., Chmelik, C., Scheidt, H., Vasenkov, S., Staudte, B., Kärger, J., Kremer, F., Zdronza, G. and Kornatowski, J., *J. Am.Chem.Soc.* **124** (2002) pp. 8690.
81. Kortunov, P., Chmelik, C., Kärger, J. Rakoczy, R. A., Ruthven, D. M., Trau, Y., Vasenkov, S. and Weitkamp, J. *Adsorption* **11** (2005) pp. 235-244.

PRESSURE SWING ADSORPTION TECHNOLOGY FOR HYDROGEN PURIFICATION - A STATUS REVIEW

SHIVAJI SIRCAR

Department of Chemical Engineering, Lehigh University, Bethlehem, Pa.,18015, U.S.A.
E-mail: shs3@lehigh.edu

Pressure Swing Adsorption (PSA) processes are designed for production of hydrogen or ammonia synthesis gas from steam methane reformer off gas with or without by-product carbon dioxide, as well as for production of H_2 from refinery off gases. A variety of adsorbents are used for these processes. The ease of desorption often dictates the adsorbent selection. Empirical PSA process performance data are used to fine- tune mathematical design models. The hydrogen productivity of the PSA process can be increased by rapid PSA process cycles. The hydrogen recovery can be increased by hybridization of the PSA unit with adsorbent membranes. Novel sorption enhanced reaction processes, based on the principles of PSA, can be designed for production of hydrogen by low temperature steam-methane refining.

1. Introduction

The current global production rate of hydrogen is about 17 trillion cubic feet per year [1]. The H_2 is used in petroleum refining, ammonia and methanol production, food industry, chemical and petrochemical industries, metal refining, electronic industry, etc. Use of H_2 as a clean fuel is also an emerging market. The advent of 'Hydrogen Economy' and 'Stricter Environmental Regulations' are continually increasing the H_2 demand [2, 3]. Pressure Swing Adsorption (PSA) has become the state of the art technology for production of high purity H_2 (99.995+ %) from a feed gas containing 60 – 90 % H_2. It is used by more than 85 % of global H_2 production facilities in the size range of 1- 130 MMSCF of H_2 per day. The trend is to build even larger single train PSA units. The two most commonly used gas sources for H_2 production are (i) Steam-Methane-Reformer Off Gas (SMROG) after it has been further treated in a water-gas-shift (WGS) reactor, and (ii) Refinery Off Gases (ROG) from various sources [4]. They are available at a pressure of 4-30 bars and a temperature of 20-40 C, and are saturated with water. The typical gas compositions (dry basis) are 70-80% H_2, 15-25% CO_2, 3-6% CH_4, 1-3% CO, and trace N_2 , and 65-90% H_2, 3-20%

CH_4, 4-8% C_2H_6, 1-3% C_3H_8, and <0.5% C_4H_{10}+ for the SMROG and the ROG, respectively.

The basic principle of a H_2 PSA process for these applications is relatively simple. The bulk and dilute impurities present in the feed gas are adsorbed by passing it through a column packed with one or more adsorbents in order to produce the pure H_2 product gas at feed pressure. The impurities are then desorbed by lowering their super-incumbent partial pressures inside the column in order to produce an impurity rich gas. The two common methods of lowering the impurity partial pressure are (i) decreasing the total column pressure (counter-current depressurization), and (ii) flowing a part of the impurity-free H_2 product gas over the adsorbent at a lower pressure (purge). Though simple in principle, a practical PSA process can be fairly complex, consisting of the adsorption and the desorption steps in conjunction with a variety of complementary steps which are designed to improve the H_2 product purity and recovery, and to reduce the adsorbent inventory [5]. Thus, a PSA process involves a series of sequential, non-isothermal, non-isobaric, unsteady- state steps operated in a cyclic steady state fashion using multiple, parallel adsorption columns.

2. Versatility of H_2 PSA Processes

The PSA technology is a very versatile gas separation tool. Many different PSA processes have been developed for purification of H_2 during the last thirty five years. The effort remains unabated. A survey shows that 275 U.S. Patents on H_2 PSA were issued during 1978- 2005 to 73 corporations around the world [6]. The following section briefly describes four different H_2 PSA processes in order to demonstrate the versatility and the design flexibility of this technology:

Poly-Bed PSA Process: The most frequently used H_2 PSA processes are designed for sole production of high purity H_2 from the feed gas. A popular design called 'Poly-bed process' was introduced by the Union Carbide Corporation and later sold to the UOP Corporation [7]. Figure 1 is a schematic flow diagram of a ten-column Poly-bed system employing nine sequential steps which are listed in the figure. The adsorbers are packed with a layer of activated carbon in the feed end and a layer of a 5A zeolite in the product end. The process was originally designed to produce high purity H_2 at high H_2 recovery from SMROG.

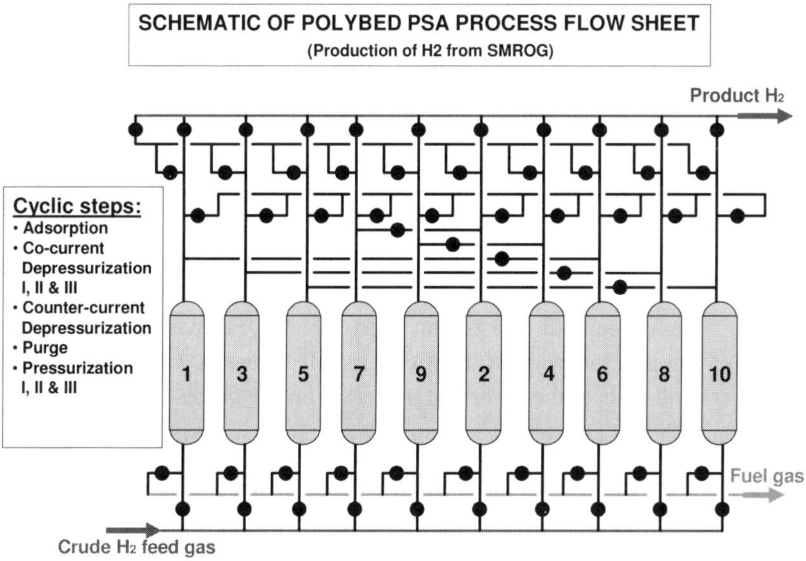

Figure 1. Schematic Drawing of Poly-Bed H2 PSA Process

A detailed description of the Poly-bed PSA process and its operation can be found elsewhere [4, 7, 8]. The unique features of this process are (i) stopping the high pressure adsorption step when the leading impurity mass transfer zone from the feed gas travels about half way through the column and the remainder of the column remains free of the impurities, (ii) co-currently depressurizing the column to a near ambient pressure level in three sequential steps in order to produce pure H_2 streams at three different pressures by adsorbing the impurities from the left-over void gas in the clean section of the column, and (iii) using these H_2 streams to counter-currently purge and pressurize some of the companion columns. This mode of operation significantly improves H_2 recovery by the PSA process by extracting H_2 from the void gas at the end of the adsorption step.

LOFIN PSA Process: A very interesting variation of the Poly-bed process was developed by Toyo Engineering Corporation [9] for production of H_2 from ROG. A flow diagram of the process using four adsorption columns is given in Figure 2 which also lists the cyclic steps for the process. The adsorbers are packed with a layer of silica gel at the feed end and a layer of activated carbon in the product end.

A detailed description of the LOFIN process can be found elsewhere [4, 9-11]. The cyclic steps of the process are very similar to those of the Poly-bed process. A unique difference is that the impurities are allowed to breakthrough the adsorption column during the co-current depressurization step which produces the H_2 gas for counter-currently purging a companion column. This effluent gas, which is initially pure H_2 and later contains some impurities, is stored in a gas storage vessel packed with an inert material. The stored gas is then used to purge an adsorber by reversing the direction of flow through the storage vessel so that the adsorber is purged first with impure H_2 and then with pure H_2. This concept of 'Last- Out First- IN (LOFIN)' provides a larger quantity of H_2 purge gas without sacrificing its effectiveness, which improves the H_2 recovery and reduces the adsorbent inventory for the process.

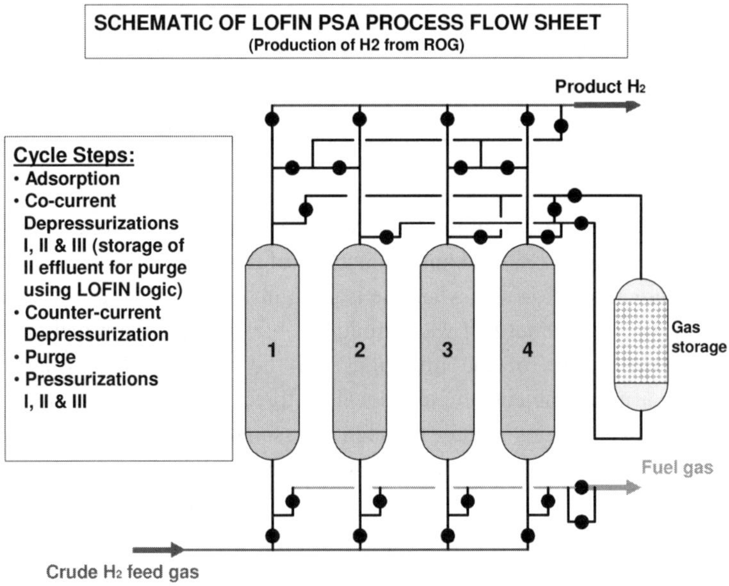

Figure 2. Schematic Drawing of LOFIN PSA H_2 Process

Gemini PSA Process: The Gemini PSA process was developed by Air Products and Chemicals Corporation for simultaneous production of high purity H_2 and CO_2 from SMROG [12]. The process uses two sets of multiple- columns (A & B) operated in series. The A columns are packed with activated carbon primarily for removal of CO_2 from the feed gas. The B columns are packed with 5A zeolite for removal of the other impurities. A detailed description of the Gemini process can be found elsewhere [4, 12, 13]. Figure 3 shows a schematic flow diagram of the process employing six A beds and three B beds and lists the cyclic steps. The unique features of the Gemini PSA process include (i) use of a CO_2 rinse step at feed pressure following the adsorption step in order to purge out the left-over void gases, which is recycled as feed, (ii) de-coupling the A and the B beds during regeneration, and (iii) using different schemes for regeneration, such as evacuation for A beds and depressurization and H_2 purge for B beds.

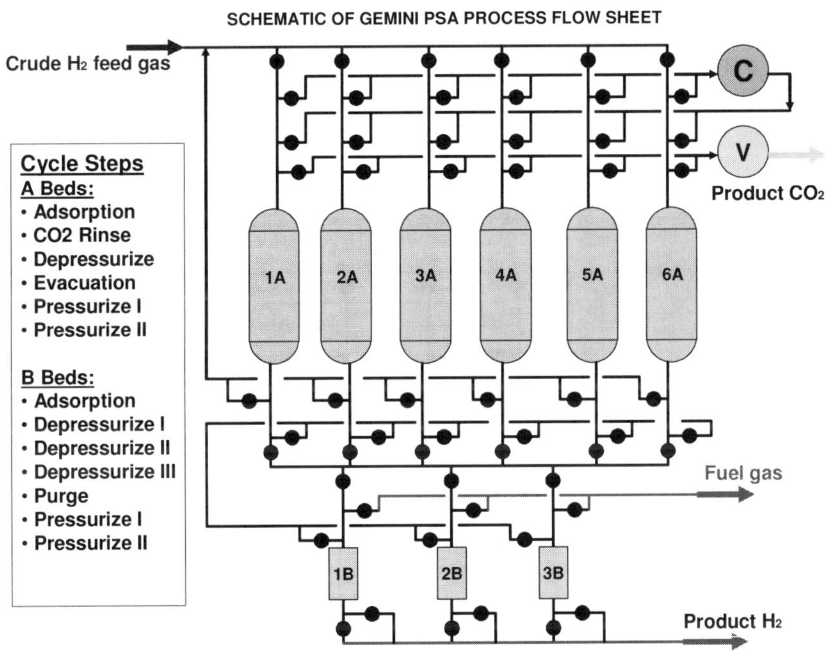

Figure 3. Schematic Drawing of Gemini PSA Process

These regeneration schemes allow simultaneous production of H_2 and CO_2 by the process with high purity and recovery for both components. Production of a CO_2 by product from SMROG is a valuable feature of the Gemini process because the CO_2 can be sequestered after necessary compression to minimize the green house gas emission to the atmosphere. The amount of CO_2 emission from a 100 MMSCFD H_2 PSA plant is ~ 1500 tons per day.

<u>Gemini – NH_3 PSA Process:</u> The above-described Gemini PSA process was modified for simultaneous production of ammonia synthesis gas (mixture of 1:3 N_2:H_2) and CO_2 from SMROG feed [14, 15]. This was achieved by purging the B beds and partially pressurizing the A and the B beds with extraneous N_2 instead of product H_2. This mode of operation introduced N_2 into the adsorbers prior the adsorption step. The weakly adsorbed N_2 was then expelled out in conjunction with the H_2 product as the ammonia synthesis gas at feed gas pressure in the subsequent adsorption step. Figure 4 is a schematic drawing of the modified Gemini PSA process employing four A beds and two B beds. It also lists the cycle steps. The elimination of the H_2 purge step results in higher H_2 recovery than the original Gemini process. However, a portion of the imported N_2 used in the process is lost as the PSA waste gas. The modified Gemini process can be very attractive for production of urea by reacting the primary and the secondary products [$2NH_3 + CO_2 \leftrightarrow NH_2.CO.NH_2 + H_2O$].

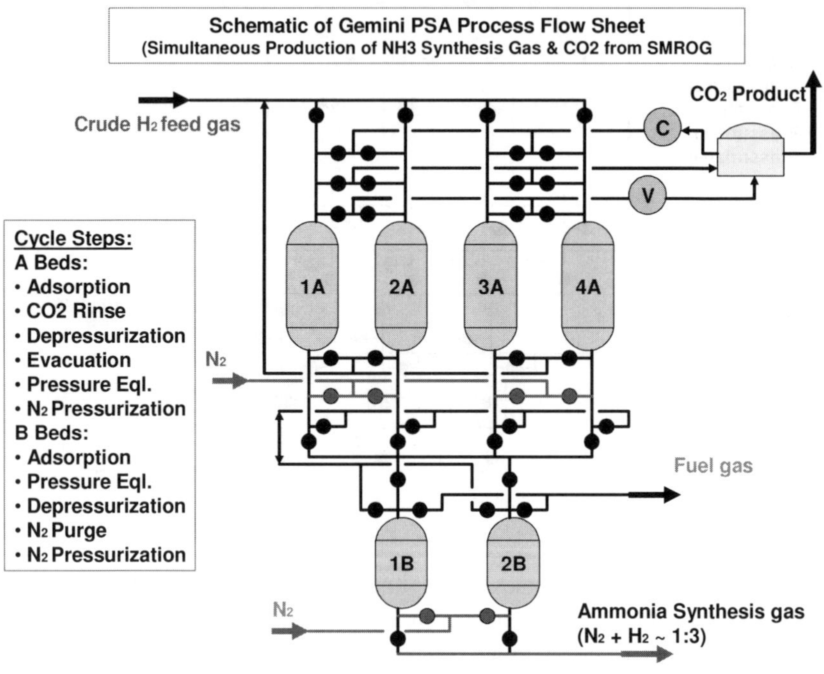

Figure 4. Schematic Flow Diagram of Gemini- NH_3 PSA Process

Examples of process performance of these four PSA processes are given in the table below [7, 9, 12, 14]. The high separation efficiency of these processes is self evident.

Table 1. Examples of process performance of these four PSA processes

Process	Feed Gas	Primary Product			Secondary Product			Ref.
		Gas	Purity	Recovery	Gas	Purity	Recovery	
Poly-bed	SMROG at 20.7 bar	H_2	99.999%	86.0%	None	--------	--------	[7]
LOFIN	ROG at 28.0 bar	H_2	99.96%	86.3%	None	--------	--------	[9]
Gemini	SMROG at 18.0 bar	H_2	99.999%	87.1%	CO_2	99.4%	94.0%	[12]
Gemini-NH_3	SMROG at 18.0 bar	N_2+ H_2	H_2 ~75% N_2 ~25%	~ 95% ~ 75%	CO_2	99.4%	94.0%	[14]

3. Adsorbents for H_2 PSA processes

Adsorbent selection is a critical issue for efficient operation of the H_2 PSA processes. The important adsorptive properties include (i) adiabatic working capacity, (ii) selectivity of adsorption, (iii) isosteric heat of adsorption, and (iv) desorption characteristics of the impurities being removed by the adsorbent. All of these properties play a role in the selection of the optimum adsorbent. However, ease of desorption of the impurity is often the controlling criterion for adsorbent selection [4]. The adsorbents chosen for practical H_2 PSA processes generally exhibit high mass transfer coefficients for the impurities and the separation is primarily governed by their thermodynamic selectivity. Several layers of different adsorbents are often used in a single adsorber. The following table lists the commonly used adsorbents for removal of the impurities present in SMROG and ROG [4]:

Table 2. Adsorbents for removal of the impurities present in SMROG and ROG

SMROG	H_2O Alumina	CO_2 Activated Carbon	CO 5A Zeolite	CH_4 Activated Carbon, 5 A Zeolite	N_2 5 A Zeolite
ROG	H_2O Alumina	CH_4 Activated Carbon	C_2H_6 Activated Carbon	C_3H_8 Silica Gel	C_4H_{10} Silica Gel

The pure gas adsorption isotherms of the components of SMROG (dry basis) on the BPL activated carbon and 5 A zeolite are shown in Figures 5 (a) and (b), respectively [4]. The polar zeolite adsorbs the polar components of SMROG (CO_2, CO and N_2) much more strongly and exhibits higher capacities for these gases at a given partial pressure than the carbon. The Henry's Law selectivity of adsorption of CO_2 over H_2 on the zeolite and the carbon are 7400 and 90.8, respectively [4]. On the other hand, the adsorption isotherms of non-polar CH_4 on both adsorbents are similar. CH_4 is selectively adsorbed over CO on the carbon and CO is selectively adsorbed over CH_4 on the zeolite [4].

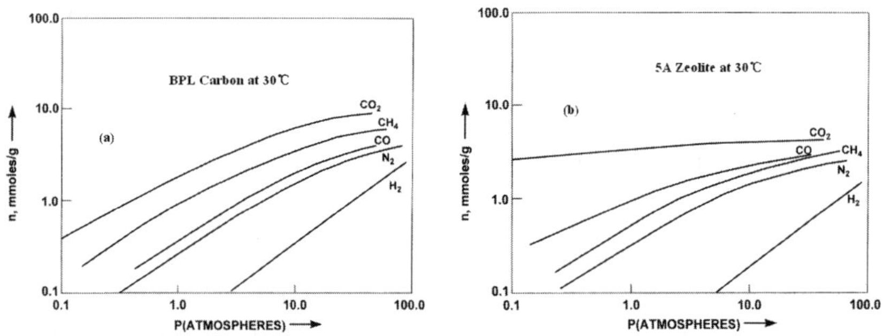

Figure 5. Adsorption Isotherms: (a) BPL Carbon, (b) 5 A Zeolite

Despite the larger capacity and selectivity of adsorption of CO_2 on the zeolite, the activated carbon is chosen as the preferred adsorbent for bulk CO_2 removal from SMROG because it is easier to desorb CO_2 from the carbon by H_2 purge as shown by Figure 6 [4]. It shows the fractional amount of CO_2 desorbed from a BPL carbon or 5A zeolite column, which was initially equilibrated with CO_2 at 1 bar and 30℃, as a function of the specific amount of H_2 leaving the column during the isobaric and isothermal purge process. Clearly, much less H_2 is consumed to remove CO_2 from the carbon column. This property makes the activated carbon the material of choice for removal of bulk CO_2 by a PSA process.

The selection of 5A zeolite for removal of dilute CO and N_2 from SMROG, on the other hand, is based on the higher working capacity and selectivity of adsorption of these gases on the zeolite than those on the carbon. The zeolite requires a larger amount of H_2 purge gas to desorb these gases than the carbon, but the amount of H_2 needed to purge out a significant fraction of the adsorbed gases is relatively small [4].

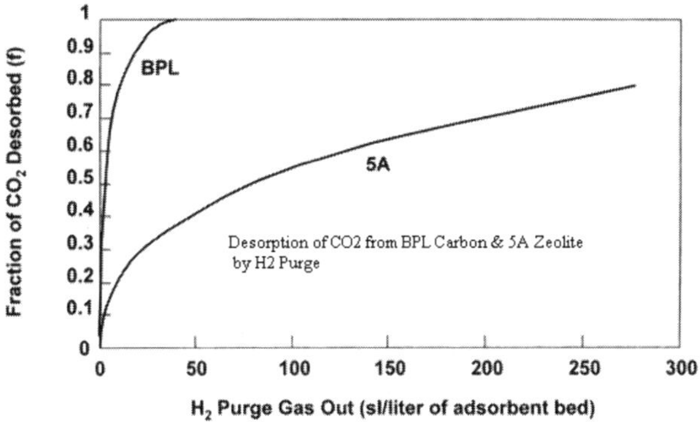

Figure 6. Desorption of CO_2 by H_2 Purge at 1 bar and 30℃

The ease of desorption of C_3+ hydrocarbons from the silica gel makes it the preferred adsorbent for production of H_2 from ROG even though the activated carbon offers larger adsorption capacity and selectivity for these gases. The carbon is chosen for removal of relatively weakly adsorbed C_1 and C_2 hydrocarbons from ROG because of its higher working capacity and selectivity of adsorption for these gases [4].

Research on developing better adsorbents for H_2 PSA applications is an on going effort. Structural and chemical modifications of activated carbons and synthesis of mixed-cation exchanged zeolite frameworks are two active areas of research [16]. Increasing impurity mass transfer coefficients into the adsorbent particles is another important goal needed for reducing the adsorption time of the PSA cycle, and thus reduce adsorbent inventory or increase H_2 productivity.

4. Recent Developments in H_2 PSA Technology

Three recent developments in the field of H_2 PSA technology are briefly described in this section. They address three very different goals.

4.1. *Rapid Pressure Swing Adsorption (RPSA) processes for H_2 purification*

Development of scaled-down versions of H_2 PSA processes producing 0.05 – 1.0 MMSCFD H_2 will be necessary for supporting the forth coming 'hydrogen economy'. They will serve numerous H_2 based applications like H_2 fuel-cells, internal combustion vehicles, stationary or portable power generators, power generators for remote locations, etc [16].

Very compact and low cost H_2 PSA units are being developed for this purpose by operating a conventional H_2 PSA cycle (total cycle time of10 -30 minutes) using a very short total cycle time (0.5 -1.5 minutes) and employing two specially designed rotary valves in place of an array of standard switch valves [16]. A Questair Corporation RPSA- H_2 unit employing 6 -9 adsorber beds and rotary valves can process a SMROG to produce a high purity H_2 gas (<1 ppm CO) with H_2 recovery of ~80% at a much higher (4-10 times) H_2 productivity than a conventional PSA unit. These units can be designed to produce 4000 SCFD to 4 MMSCFD of H_2 [17].

A few inherent limitations of a RPSA process are that (i) the short cycle time prevents incorporation of all of the process steps of a conventional PSA cycle which improve separation efficiency, (ii) the productivity (lb moles of product/lb of adsorbent/time) of the process can not be increased indefinitely by lowering the cycle time, there being a finite limiting value of productivity for a finite value of the adsorbate mass transfer coefficient [18], and (iii) instantaneous thermal equilibrium between the gas and the solid adsorbent inside an adsorber can not be achieved when the cycle times are very short, which will adversely affect the working capacity of the adsorbent [19]. The last two findings were demonstrated by a simplified analysis of idealized PSA processes on a single adsorbent particle. Nevertheless, the development of rapid PSA processes opens up further research and development opportunities on (i) novel adsorbent configurations such as structured adsorbents, and (ii) innovative mechanical devices for operating the rapid cycles.

4.2. *Sorption Enhanced Reaction Process (SERP) for production of H_2*

Catalytic steam-methane reforming (SMR) is the popular commercial method of H_2 production. Figure 7 shows a flow diagram of this route of H_2 production consisting of a SMR reactor, a WGS reactor, a PSA H_2 purification unit, and heat exchangers for heat recovery [20].

The over-all equilibrium-controlled SMR reaction (CH4 + 2H2O ↔ CO2 + 4H2) is highly endothermic, and the reactor is operated at a very high temperature of ~ 850 ℃to get a decent conversion of CH_4 to H_2. This requires that the reactors be made from expensive alloyed steel. The SERP concept simultaneously carries out the SMR reaction and the H_2 purification process in a single unit operation. Furthermore, the reaction is carried out at a much lower temperature (~ 400 -500℃) without sacrificing the conversion of CH_4 to H_2. Thus the reactors can be made from ordinary steel.

The concept is based on Le Chatelier's principle that removal of an undesired reaction product from the reaction zone of an equilibrium- controlled reaction increases the conversion and the rate of formation of the desired component. The process uses a sorber-reactor which is packed with a physical admixture of a reforming (noble metal on alumina) catalyst and a chemisorbent (K_2CO_3 promoted hydrotalcite), which selectively and reversibly chemisorbs CO_2 from the gas phase of the reaction zone at a temperature of ~ 450°C in presence of steam. The chemisorbent is periodically regenerated by using steam purge under vacuum so that it can be re-used in a cyclic manner using the principles of PSA [21]. Figure 7 shows the flow diagram of a two-column embodiment of the SERP concept. It also lists the cyclic process steps of the process. Table 3 gives an example of the performance of the SERP concept for direct production of fuel-cell grade H_2 and compares that with the corresponding performance of a conventional SMR reactor [22]. The compactness and the advantages of the SERP concept are obvious.

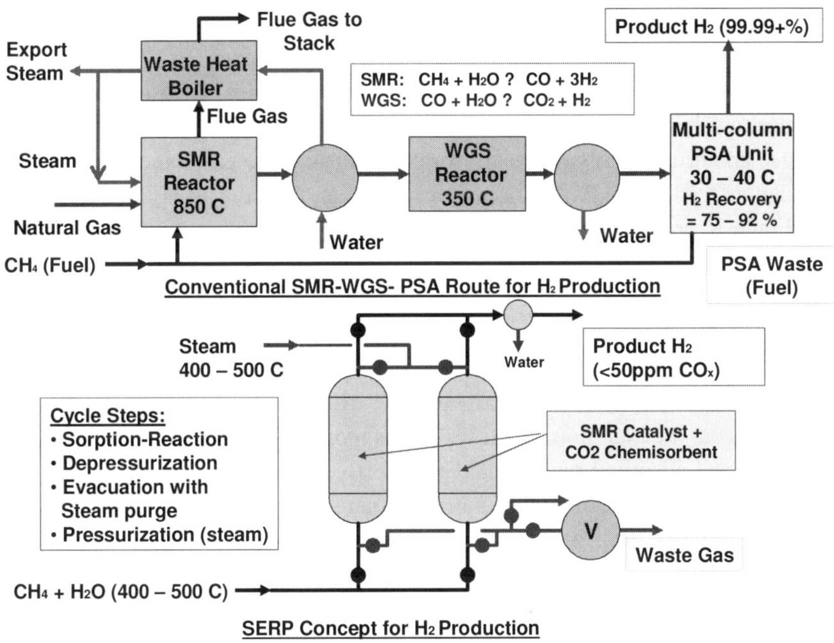

Figure 7. Flow Diagrams for the conventional SMR and SERP Concepts

Table 3. Gives an example of the performance of the SERP concept for direct production of fuel-cell grade H_2

Process	Product Purity (Dry Basis), mole %				CH_4 to H_2
Feed gas: 6: 1 H_2O: CH_4	H_2	CH_4	CO_2	CO	Conversion, %
T = 490℃, P = 11.4 psig					
SERP Concept	94.4	5.6	40 ppm	30 ppm	73.0
Conventional SMR Reactor	67.2	15.7	15.9	1.2	52.6

4.3. *Hybrid adsorbent membrane – PSA process for improving H_2 recovery*

The recent increase in the price of natural gas and the growth in H_2 demand has put a premium on improving the over-all H_2 recovery from SMROG. One approach to achieve that goal is to recover a part of the H_2 from the PSA waste gas (Figure 7) containing 30-40 % H_2.

Integration of a H_2 PSA process with an adsorbent membrane can meet this goal [23, 24]. A nano-porous carbon adsorbent membrane called 'Selective Surface Flow (SSF)' membrane which selectively permeates CO_2, CO and CH_4 from their mixtures with H_2 by an adsorption- surface diffusion-desorption transport mechanism may be employed for this purpose. The SSF membrane can produce an enriched H_2 gas stream from a H_2 PSA waste gas, which can then be recycled as feed to the PSA process for increasing the over-all H_2 recovery. The membrane is prepared by controlled carbonization of poly-vinyledene chloride supported on a macro-porous alumina tube. The membrane pore diameters are between 6 -7 A, and its thickness is ~ 1-2 μm [25].

Figure 8a shows a cartoon of the transport mechanism through the SSF membrane. Larger and more polar molecules (CO_2, CO and CH_4) are selectively adsorbed on the pore walls of the membrane over the smaller molecules (H_2) of the feed gas at the high pressure side. CO_2 is more selectively adsorbed than CO and CH_4. The adsorbed molecules then selectively diffuse on the pore walls to the low pressure side of the membrane where they desorb producing a CO_2 enriched permeate gas. A H_2 enriched gas is produced at feed pressure as the primary product.

Furthermore, the membrane can efficiently operate (high selectivity and flux) under a moderate pressure gradient across the membrane. These are some of the unique features of the SSF membrane.

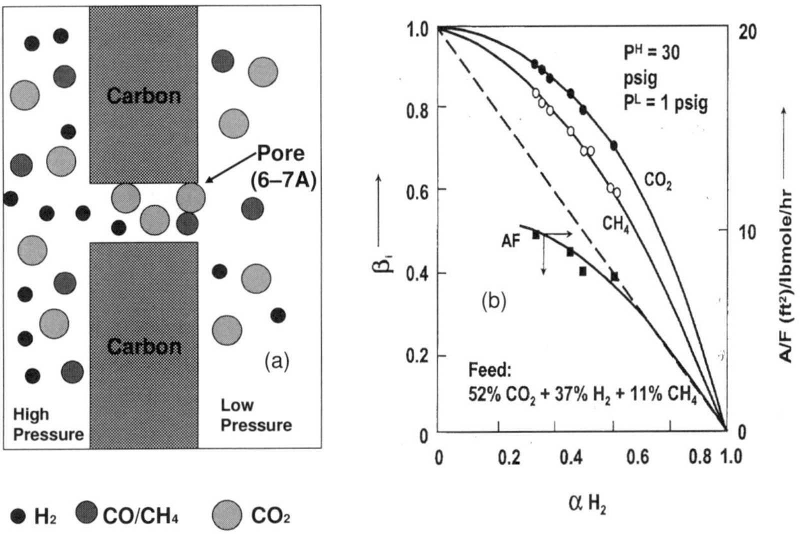

Figure 8. (a) Transport mechanism and (b) Performance of SSF membrane

Figure 8b depicts the performance of a SSF membrane for a feed gas which is representative of a H_2 PSA waste gas [23]. The pressures in the high and the low pressure sides of the membrane are ~3 and 1 bars, respectively. The figure plots rejection of component i (β_i) of the feed gas and the membrane area needed to process a given feed gas flow rate (A) as functions of H_2 recovery (α_{H2}). About 90% CO_2 and 80% (CH_4+ CO) can be rejected when the H_2 recovery is 40%.

Figure 9 shows a schematic flow diagram and an example of the hybrid H_2 PSA-SSF membrane concept. The fresh feed to the PSA process is SMROG. The PSA process cycle is an abridged version of the Poly-bed process with only two co-current depressurization steps, having a H_2 recovery of 77.6%. The countercurrent depressurization effluent gas is fractionated. The initial part of this gas, which is richer in H_2, is directly fed to a SSF membrane at a pressure of 3 bar. The H_2 purge effluent gas is compressed to 3 bar and fed to the same membrane. The H_2 enriched high pressure effluent gas from the membrane is recompressed and recycled as feed gas to the PSA process. This increased the overall H_2 recovery of the hybrid process to 84.0% [23].

The SSF membrane can also be used to enrich H_2 from the waste gas of a PSA process purifying the ROG because it selectively permeates C_1- C_4 hydrocarbons from mixtures with H_2 [26]. The membrane is particularly

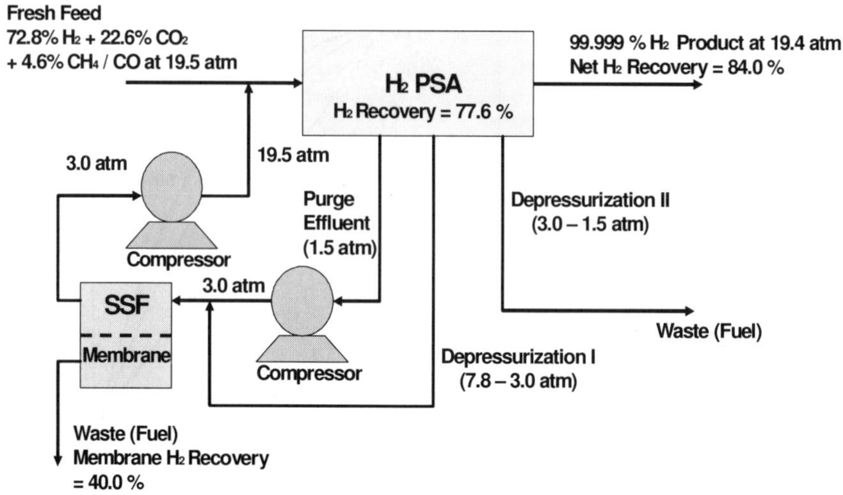

Figure 9. Schematic flow sheet of a hybrid H_2 PSA-SSF membrane process

effective in removing C_2+ hydrocarbons from H_2. Consequently, it can also be integrated with a PSA unit purifying H_2 from ROG in order to increase the over-all H_2 recovery.

5. Engineering Design of H_2 PSA Processes

The design requirements for an industrial H_2 PSA process can be very stringent. The H_2 product purity must be 99.995 mole% or better for most applications. At the same time, an error of ± 2 percentage points in the estimation of the H_2 recovery can make or break the economics of a process design [27].

It may not be possible to theoretically design a H_2 PSA process with such accuracy without using the actual experimental process performance data to fine tune the design model. The reasons are that (i) the practical PSA processes are fairly complex and (ii) the key input data (multi-component adsorption equilibria, kinetics and isosteric heats) for the mathematical design model (integration of coupled partial differential equations describing the mass, the heat, and the momentum balances inside the adsorber) may not be very accurate [27]. The PSA process models often act as amplifiers of errors in the input data.

Consequently, the commercial design and optimization of a H_2 PSA process still largely remains an empirical effort. The process simulation models are, however, extremely valuable for screening new ideas and adsorbents, parametric study of the processes for optimization, establishing process limitations, process scale-up, and design of control schemes. The models are often modified using actual H_2 PSA plant performance data so that they can be used as reliable design tools. Corporations designing and selling H_2 PSA systems develop their own proprietary PSA process models and database.

There are very few publications which compare simulated H_2 PSA process performance using multi-component, non-isothermal models with those obtained experimentally, particularly for production of high purity H_2 from SMROG or ROG- like feeds [28- 32]. Figures 10a and b show two examples. The solid and the dashed lines are the simulation results using adiabatic and isothermal columns, respectively. The points are experimental data. The ROG feed was purified using a six bed system packed with a layer of silica gel and a layer of activated carbon [31]. The SMROG feed was purified with a four bed system packed with a layer of an activated carbon and a layer of 5A zeolite [32]. The cycle steps for both systems were similar to those of the Poly-bed PSA process.

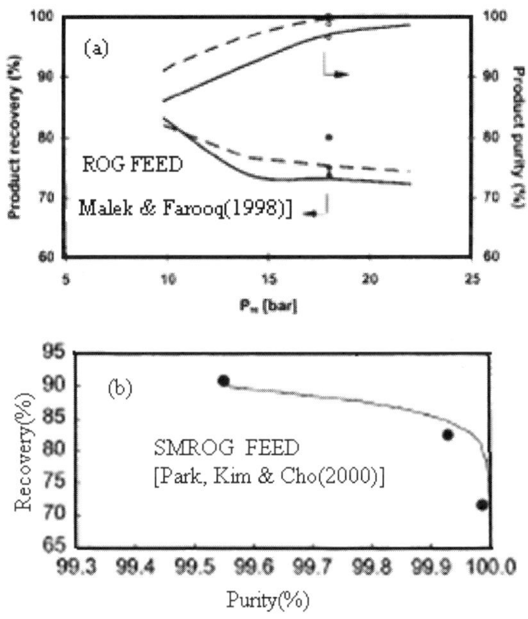

Figure 10. Comparison between H_2 PSA model performance and experiment

The Figures show that the model calculations describe the experimental performance data fairly well but the accuracy needed by industrial design may still be lacking.

6. Summary

PSA is the state of the art technology for production of high purity H_2 from SMROG and ROG. PSA processes are also available for simultaneous production of H_2 or NH_3 synthesis gas and CO_2 from SMROG. Different adsorbents including activated carbons, zeolites, silica gels and aluminas are used in H_2 PSA processes. Ease of desorption often dictates adsorbent selection. Packing adsorbers with layers of different adsorbents is a common practice.

Design of rapid H_2 PSA cycles using rotary valves to enhance the H_2 productivity and to reduce the plant foot print is a trend. Other emerging ideas include (i) sorption enhanced reaction concept for low temperature production of fuel-cell grade H_2 by SMR which employs a CO_2 chemisorbent and a novel PSA scheme, and (ii) hybrid adsorbent membrane – H_2 PSA systems for increasing the over-all H_2 recovery from the feed gas. Mathematical models for design of H_2 PSA processes are very useful for process optimization, adsorbent screening, establishing process limitations, etc. Experimental process data may be needed to fine tune the models for use as a practical design tool.

References

1. Suresh, B, Schlag, S, Inoguchi, Y, "Hydrogen", CEH Marketing Research Report, (2004).
2. "The Hydrogen Economy", National Academy Press, Washington, D. C. (2004).
3. Towler, G. P, Mann, R, Serriere, J. L, Gabaude, M. D, I &E C. Res., 35, 2378 (1996).
4. Sircar, S, Golden, T. C, Sep. Sci. Technol., 35, 667 (2000).
5. Sircar, S, "PSA Technology" in Adsorption Sci. Tech., A. E. Rodrigues et al (eds), Kluwer Academic Publishers, Dordrecht, The Netherlands, pp 285-321 (1989).
6. World Patent Index, Derwent Publication, London.
7. Fuderer, A, Rudelstorfer, E, U.S. Patent 3,986,849 (1976).
8. Stocker, J, Whysall, M, Miller, G. Q, "30 Years of PSA Technology for Hydrogen Purification", UOP LLC web-site, Des Plaines, IL., (1998).
9. Yamaguchi, T, Kobayashi, Y, U.S. Patent 5,250,088 (1993)
10. Yamaguchi, Ohkamo, U, Kobayashi, Y, "Hydrogen Recovery & Purification by LOFIN Process", Proceedings of 10th World Hydrogen Conf., Block, D. L and Nejat, V. T. (eds), Published by Florida Solar Energy Cent., 3, 1497 (1994).
11. Okama, U, "Increased H_2 Recovery with Advanced PSA Technology", PTQ Summer, pp95-98 (1996).
12. Sircar, S, U.S. Patent 4,171,206 (1979).

13. Sircar, S, Kratz, W. C, Sep. Sci. Tech., 23, 2397 (1988).
14. Sircar, S, U.S. Patent 4,375,363 (1983).
15. Sircar, S, Sep. Sci. Tech., 25, 1087 (1990).
16. Sircar, S, Golden, T. C, "Pressure Swing Adsorption Technology for H_2 Production", Chapter 12 in "Hydrogen Production Technologies", Liu, K, Song, C, Velu, S (eds), Wiley Inter-science Publication, in press (2006).
17. Private communication with Questair, Inc., Canada
18. Sircar, S, Hanley, B. F, Adsorption 1, 313 (1995).
19. Sircar, S, Adsorption 11, 509 (2005).
20. Leiby, S. M, "Options for Refinery H_2", PEP Report # 212, Process Economics Program, SRI International, Menlo Park, CA (1994).
21. Hufton, J. R, Mayorga, S, Sircar, S., AIChE J., 45, 248 (1999).
22. Waldron, W. E, Hufton, J. R, Sircar, S, AIChE J., 47, 1477 (2001).
23. Sircar, S, Waldron, W. E, Rao, M. B, Anand, M, Gas Sep. Purif., 17, 11 (1999).
24. Sircar, S, Waldron, W. E, Anand, M, Rao, M. B, U.S. Patent 5,753,010 (1998).
25. Rao. M. B, Sircar, S, J. Membrane Sci., 110, 109 (1996).
26. Naheiri, T, Ludwig, K. A, Anand, M, Rao, M. B, Sircar, S, Sep. Sci. Technol., 32, 1589 (1997).
27. Sircar, S, "Basic Research Needs for Design of Adsorptive Gas Separation Processes", I & E C. Res. in press (2006).
28. Cen, P, Yang, R. T, Sep. Sci. & Tech., 20, 725 (1985).
29. Warmuzinski, K, Tanczyk, M, Chemical Engineering & Processing, 36, 89 (1997).
30. Zhu, D, Tianranqi Huagong (in Chinese), 23, 36 (1998).
31. Malek, A, Farooq, S, AIChE J., 44, 1985 (1998).
32. Park, J. H, Kim. J. N, Cho, S. H, AIChE J., 46, 790 (2000).

NEW NANOPOROUS ADSORBENTS

A. KONDO, Y. TAO, H. NOGUCHI, S. UTSUMI, L. SONG, T. OHBA, H. TANAKA, Y.HATTORI, T. ITOH, H. KANOH, C. M. YANG, M. YUDASAKA[*], S. IIJIMA[*,**] AND K. KANEKO

Nanoscale Science, Graduate School of Science and Technology, Chiba University, Yayoi 1-33, Inage, Chiba 263-8522, Japan

[*]*JST/SORST, NEC Corporation, 34 Miyukigaoka, Tsukuba 305-8501, Japan*

[**]*Department of Physics, Meijo University, 1-501 Shiogamaguchi, Tenpaku, Nagoya 468-8502, Japan*

New trials to improve adsorption kinetics of zeolites and activated carbon fiber(ACF)s with addition of mesopores with the aid of templating and chemical modification are described. The templating with carbon aerogel and resorcinol-formaldehyde gels added mesopores of 10-12 nm in width to ZSM-5, NaA, and NaY. The steam reactivation of ACF with $Ca(NO_3)_2$ provided mesopore-added ACF, whose adsorption rate for methylene blue was remarkably improved. The clathrate compound formation mechanism of metal organic framework of copper with CH_4 and CO_2 was shown for gate adsorption that induces predominant adsorption and desorption at the definite pressures. The adsorption of H_2 and D_2 on single wall carbon nanohorn (SWNH) was examined over the temperature range of 20 K to 77 K. The adsorption amount of D_2 was larger than that of H_2, which was explained by the quantum molecular sieving effect. Other adsorption abilities of SWNH assemblies were described.

1. Introduction

The urgent demand for preservation of the global environments has requested to construct environment-friendly technologies. Adsorption has contributed to energy storage, highly efficient catalysis, concentration of noble substances, and removal of pollutants, separation of harmful gases or valuable gases, medical treatments, forming the principal bases of various technologies. Therefore, development of adsorption science and technology is clue to support a peaceful and pleasant human society. One of the important issues on adsorption science and technology is supplying optimum adsorbents for environment-friendly chemical processes. Then, many nanoporous adsorbents have been developed as hopeful adsorbent applicants of high specificity and efficiency. Zeolites and activated carbons have been widely used in various technologies. Even these conventional adsorbents need better adsorption characteristics. As pore width of zeolites and activated carbons are less than 2 nm (typical micropores according to the IUPAC classification), adsorption of large molecules is often perturbed due to the diffusion restriction in the micropores. Hence, addition of mesopores has been required to improve their adsorption kinetics and catalytic reaction activity. Mesoporous silica of the regular pore structures such as MCM and FSM have been tried to create new adsorption processes [1,2]. At the same time, nanoporous carbons of regular pore structures have been prepared using the templating of mesoporous silica [3]. Then, the templating synthesis has become a major route to prepare the designed nanoporous solids. This article introduces two examples of mesopore-added zeolites with templating method and mesopore-added activated carbon fibers (ACFs).

Organic chemistry and coordination chemistry are going to provide new types of nanoporous solids, so called metal organic frameworks (MOFs) or organic-inorganic hybrid crystals. The MOFs have soft frameworks offering nanopores of variable pore width, although they are not necessarily thermally stable. Many MOFs have been proposed as storage materials for CH_4 and H_2 [4,5], although they are not sufficient yet. This paper describes novel MOF having a unique adsorption function for CH_4 and CO_2 [6-10].

The representatives of new nanoporous materials are nanocarbons such as single wall carbon nanotube (SWNT), double wall carbon nanotube (DWNT), and multi wall carbon nanotube (MWNT). Especially an intensive expectation of nanocarbons for hydrogen storage has stimulated the adsorption studies [11,12]. The presence of impurities and erroneous evaluation of hydrogen adsorption have intervened an exact understanding of the hydrogen adsorptivity of nanocarbons. Fortunately highly pure SWNT and DWNT of several hundreds mg have been prepared very recently and thereby these samples will be available for adsorption researches soon. Nevertheless, still the amount of highly pure nanocarbons is limited. On the other hand, Iijima et al developed single wall carbon nanohorns (SWNHs) of sufficient amounts with laser ablation from pure graphite without any catalyst, which consists of single graphen wall [13]. Furthermore, nanoscale holes (nanowindows) can be added on the wall of SWNH, giving rise to a remarkable molecular sieving effect [14]. This paper describes the nanoporosity and quantum molecular sieving effect for H_2 and D_2.

2. Mesopore-added zeolite and activated carbon fiber

2.1. *Mesopore-added zeolite*

Zeolites are representative microporous solids of which pore width is less than 2 nm. Addition of mesopores to zeolites have been tried to improve their catalytic activity using leaching and templating techniques [15,16]. Ordinarily the templating method is hopeful to obtain zeolites having uniform mesopores irrespective of no established templating method. Authors applied carbon aerogels and resorcinol-formaldehyde (RF) gels, the precursor of the carbon aerogels, to preparation of ZSM-5, NaY, and NaA having mesopores [17-19]. It is well-known that carbon aerogels are representative mesoporous carbons, although micropores can be added [20]. These zeolites were synthesized together with carbon aerogels or RF aerogels in the mesopore channels of the templates. The templates such as carbon aerogels or RF aerogels were removed by gasification at 823 K for 18 h.

The scanning electron microscopic observation gave the presence of considerably uniform mesopores on ZSM-5, NaY, and NaA crystals, although these pores have no periodical structures. The crystalline state was guaranteed by the sharp peaks of their X-ray diffraction patterns; the peaks were slightly broader than those of the reference zeolites. Figure 1 provides clear evidences on the addition of mesopores to ZSM-5, NaY, and NaA. For example, the N_2 adsorption isotherm of ZSM-5 at 77 K overlaps with that of the mesopore-added sample below $P/P_0 = 0.4$, indicating that both zeolites have the same micropore structures. On the other hand, the mesopore-added ZSM-5 has an explicit uptake around $P/P_0 = 0.8$ with the adsorption hysteresis, showing the presence of considerably uniform mesopores. Similar results were obtained for NaY and NaA.

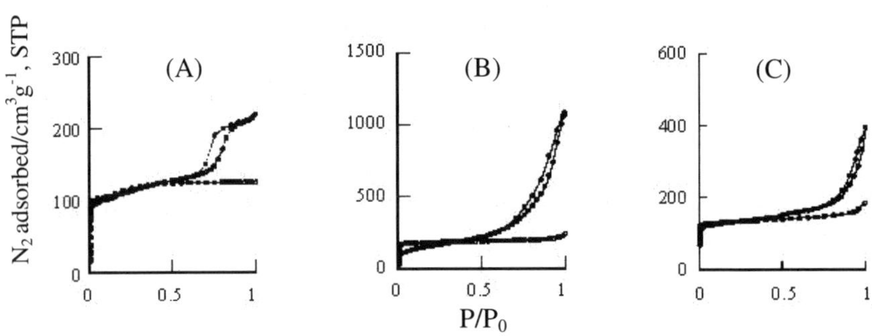

Figure 1. The N_2 adsorption isotherms of mesopore-added zeolites and zeolites without mesopores at 77 K. (A) ZSM-5, (B) NaY, and (C) NaA.

However, the overlapping below $P/P_0 = 0.4$ were not perfectly as observed in ZSM-5. These N_2 adsorption isotherms were analyzed with Dollimore-Heal (DH) method to determine the mesopore size distributions, which are shown in Fig.2. The micropore size distributions of the mesopore-added zeolites coincided with those of the reference zeolites. The mesopore size distributions are considerably uniform and their peaks are in the range of 10 to 12 nm. In particular, mesopore-added ZSM-5 gives the very sharp distribution. These mesopore-added zeolites are hopeful adsorbents and catalysts.

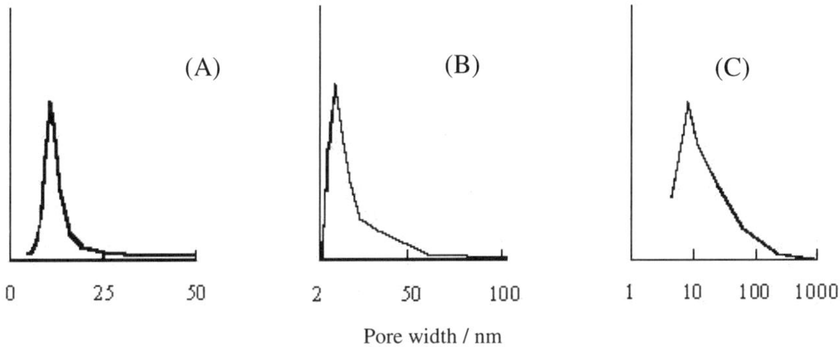

Figure 2. The mesopore size distributions of mesopore added zeolites. (A) ZSM-5, (B) NaY, and (C) NaA.

2.2. *Mesopore-added activated carbon fiber*

Activated carbon fibers (ACFs) are highly microporous carbon, which exhibit better adsorption performance than conventional granulated activated carbons due to larger external surface area of ACFs. If we can add efficiently mesopore channels to ACFs, their adsorption kinetics can be greatly improved for adsorption of large molecules such as dye molecules.Pitch-based ACF of different pore widths were reactivated with steam at 1123 K with the aid of $Ca(NO_3)$ deposition [21,22]. This reactivation could add mesopores efficiently to ACFs. Figure 3 shows the effect of mesoporosity on the adsorption rate of methylene blue (MB) on the ACFs of which micropore width is 0.7 nm. The initial adsorption rate increases greatly by addition of mesopores, because the micropore diffusion is obstacled by the MB molecules precedingly adsorbed (the molecular geometry of MB is 0.40 nm x 0.61 nm x 1.43 nm). Thus, the coexistent mesopores improve remarkably the adsorption kinetics for large molecules.

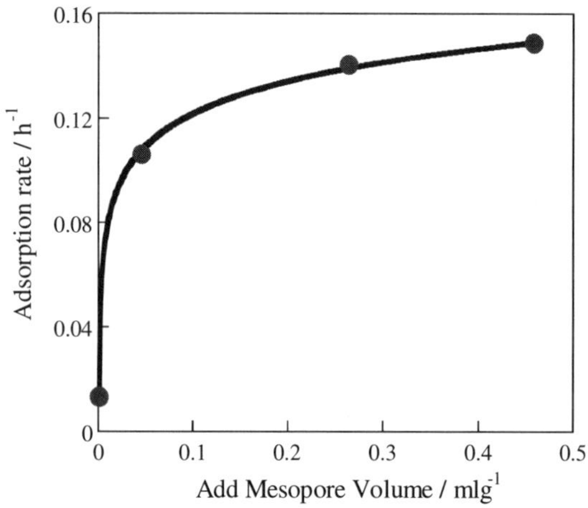

Figure 3. Effect of mesoporosity on the adsorption rate of methylene blue on ACFs.

3. Metal organic framework of gate adsorption

Active studies on gas adsorption on metal organic frameworks (MOFs) have been carried out. Li and Kaneko found new type of adsorption of CO_2 on Cu-complex crystals which have no open porosity crystallographically [6]. Hence, the compound of Cu-complex crystals is noted the latent porous crystal (LPC). Figure 4 shows the vertical adsorption and desorption isotherms of CH_4 at 273 K. We named the vertical adsorption gate adsorption. Gate adsorption behaviors were observed for CO_2, Ar, and N_2. The adsorption and desorption sensitively depends on the gas pressure and thereby the gate behavior can be applied to a new type of gas separation. The absolute adsorption capacity of CH_4 on the LPC is considerably great, because the possible volume ratio for CH_4 adsorbed is 180 vol.% at 273 K which is comparable to the DOE target value (180 vol.% at 3.5 MPa and 298 K). The temperature dependence of CH_4 adsorption indicated the clathrate formation with LPC [9]. That is, the gate adsorption is not a representative physical adsorption which does not vary the structures of both of molecules and porous solids. The in situ X-ray diffraction on CO_2 adsorption indicated the change of the unit cell structure, which is supported by the dynamic grand canonical Monte Carlo simulation for N_2 adsorption on LPC [8,10].

Figure 5 shows the relationship between the c-axis expansion and the adsorption amount from the GCMC simulation for N_2 adsorption at 77 K. The GCMC simulation indicates the step-wise adsorption, suggesting the c-axis expansion, which agrees with the experimental adsorption isotherm in Fig.5 (B). The more detailed study on the structural changes is going on. Also similar gate adsorption was observed for new MOF crystals. One MOF crystals showed double jump in N_2 adsorption isotherm at 77 K. First jump stems from micropore filling and second one is ascribed to the increase of micropore volume accompanied by a structural change [23].

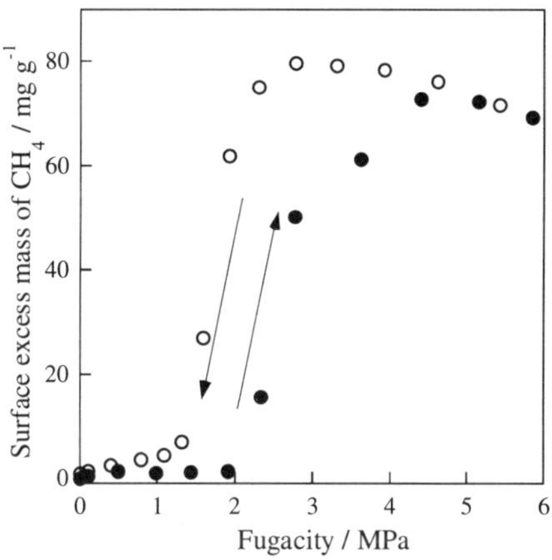

Figure 4. The adsorption isotherms of supercritical CH_4 on LPC at 273 K.

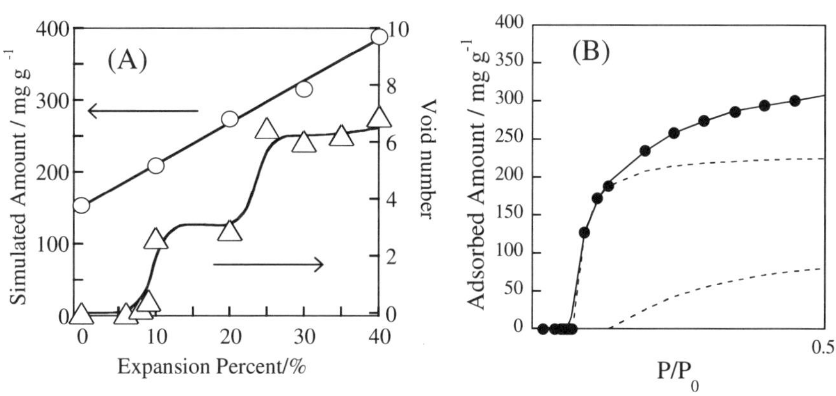

Figure 5. Changes in adsorption amount and pore volume with c-axis expansion from simulation (A) and the experimental N_2 adsorption isotherm of two stage processes (B).

4. Adsorption properties of SWNH assemblies

The nanowindows can be added to the graphene wall of SWNH by oxidation with O_2 [24]; the control of the oxidation temperature varies the nanowindow size. The nanowindow-donated SWNH shows molecular sieving property. Recently Tanaka et al have studied adsorption of H_2 and D_2 on SWNH assemblies at low temperature [25]. The thermal de Broglie wave lengths of H_2 and D_2 molecules are 0.5 nm and 0.3 nm at 20 K and 0.25 nm and 0.20 nm at 77 K, respectively. Consequently the uncertainty of the molecular position induces a marked quantum behavior depending on the mass of the molecule and the temperature. The adsorption isotherms of H_2 and D_2 on SWNH assemblies were measured over the temperature range of 20 K (boiling temperature of H_2) to 77 K. Figure 6 shows adsorption isotherms of H_2 and D_2 on SWNH assemblies without nanowindows at 20 K, 50 K, and 77 K. The lower the adsorption temperature, the greater the adsorption amount. The adsorption amount of D_2 is larger than that of H_2 at all temperatures. As the effective exclusion volume of the heavier molecule of D_2 is smaller than that of H_2, more D_2 molecules can be adsorbed in the interstitial pores of SWNH assemblies than H_2 molecules. The quantum molecular sieving effects can be interpreted by the quantum GCMC simulation with Feynman-Hibbs approximation.

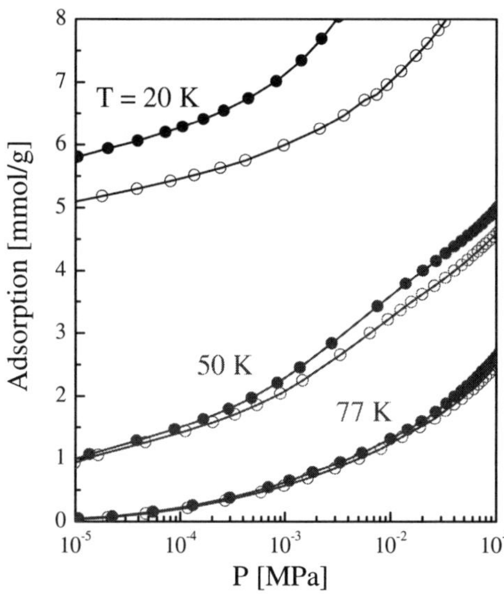

Figure 6. Adsorption isotherms of H_2 and D_2 on close SWNH assemblies at 20K, 50 K, and 77 K. Solid and open symbols denote D_2 and H_2 adsorption data, respectively.

As SWNH assemblies have single wall structures, they are hopeful adsorbents; they can provide superhigh surface area and nanopores structures. The oxidized SWNH assemblies show an excellent adsorptivity for supercritical CH_4 by compression- and chemical treatments [26-28]. Also magnetic scanning ability was donated to SWNH assemblies by doping nanoscale magnetites, which have a possibility for a medical application [29]. SWNH assemblies have characteristic n-type semiconductivity, showing a weak chemisorption responses for O_2, CO_2, and alcohols [30].

5. Future direction

This paper describes recent progresses on a part of developments and improvements on nanoporous solids. Challenges for development of new nanoporous adsorbents are indispensable to sustainable science and technology. Adsorption science and technology must take into account rapid progresses in nanoporous adsorbents. Even careful adsorption studies on highly pure SWNT and DWNT are going on in our group, suggesting inherent features of nanocarbons for adsorption science and technology near future. At the same time, we do not have sufficient understanding the fundamentals of adsorption on water and O_2, although they are very important in various technologies. We have proposed the fundamental mechanism of water on hydrophobic carbon nanopores in recent research activities [31,32].

Acknowledgement

This work was partially funded by a Grand-in-Aid for Fundamental Scientific Research (S) (no. 15101003) from the Japanese Government and by the Advanced Nanocarbon Application Project, NEDO, and Hydrogen Storage Evaluation Project, NEDO.

References

1. Kresge C. T., Leonowicz M. E., Roth W. J., Vartuli J. C.and Beck J. S., Ordered mesoporous molecular sieves synthesized by a liquidcrystal template mechanism. *Nature* **359** (1992) pp. 710-712.
2. Inagaki S., Fukushima Y.and Kuroda K., Synthesis of highly ordered Mesoporous materials from a layered polysilicate. *J. Chem. Soc., Chem. Commun.* (1993) pp. 680-682.

3. Jun S., Joo S. H., Ryoo R., Kruk M., Jaroniec M., Liu Z., Ohsuna T. and Terasaki O., Synthesis of New, Nanoporous Carbon with Hexagonally Ordered Mesostructure. *J. Am. Chem. Soc.* **122**(2000) pp. 10712-10713.
4. Kitaura R., Seki K., Akiyama G. and Kitagawa S., Porous coordination-polymer crystals with gate channels specific for supercritical gases. *Angew. Chem. Int. Ed.* **42**(2003) pp. 428-431.
5. Sudik A. C., Millward A. R., Ockwig N. W., Côté A. P., Kim J. and Yaghi O. M., Design, Synthesis, Structure, and Gas (N_2, Ar, CO_2, CH_4, and H_2) Sorption properties of porous Metal-Organic ttetrahedral heterocuboidal polyhedra. *J. Am. Chem. Soc.* **127**(2005) pp. 7110-7118.
6. Li D. and Kaneko K., Hydrogen bon-regulated microporous nature of copper complex-assembled microcrystals. *Chem. Phys. Lett.* **335** (2001) pp. 50-56.
7. Onishi S., Ohmori T., Ohkubo T., Noguchi H., Li D., Hanzawa Y., Kanoh H. and Kaneko K., Hydrogen-bond change-associated gas adsorption in inorganic-organic hybrid microporous crystals. *Appl. Surf. Sci.* **196** (2002) pp. 81-88.
8. Kondo A., Hattori Y., Kajiro H., Noguchi H., Todoh A., Tanaka H., Kanoh H. and Kaneko K., Structural change of organic complex solids upon gas adsorption. Proc. Int. Symp. Super-Functionality Organic Device IPAP Conf. Series 6 pp. 88-90.
9. Noguchi H., Kondoh A., Hattori Y., Kanoh H., Kajiro H. and Kaneko K., Clathrate-formation mediated adsorption of methane on Cu-complex crystals. *J. Phys. Chem. B* **109** (2005) pp. 13851-13853.
10. Ohba T., Inaguma Y., Kondo A., Kanoh H., Noguchi H., Gubbins K. E., Kajiro H. and Kaneko K., GCMC simulation of ynamic structural change of Cu-organic crystals with N_2 adsorption. *J. Exp. Nanosci.* **1** (2006) pp. 91-95.
11. Dillon A. C., Jones K. M., Bekkedahl T. A., Klang C. H., Bethune D. S. and Heben M. J., Storage of hydrogen in single-walled carbon nanotubes. *Nature* **386** (1997) pp. 377-379.
12. Cheng H. M., Yang Q. H. and Lui C., Hydrogen storage in carbon nanotubes. *Carbon* **39** (2001) pp. 1447-1454.
13. Iijima S., Yudasaka M., Yamada R., Bandow S., Suenaga K., Kokai F. and Takahashi K., Nano-aggregates of single-walled graphitic carbon nano-horns. *Chem. Phys. Lett.* **309** (1999) pp. 165-170.
14. Murata K., Kasuya D., Yudasaka M., Iijima S. and Kaneko K., Nanowindow-Induced Molecular Sieving Effect in Single-Wall Carbon Nanohorn. *J. Phys. Chem. B* **106** (2002) pp. 12668-12669.
15. Jacobsen C. J. H., Houzyicka C., Schmidt I. and Carlsson A., Mesoporous zeolite single crystals. *J. Am. Chem. Soc.* **122** (2000) pp. 7116-7117.

16. Tao Y., Kanoh H. and Kaneko K., Mesopore-added zeolites: An overview of their preparation, characterization and evaluation of the application. *Chem. Rev.* **106** (2006) pp. 896-910.
17. Tao Y., Kanoh H., Kaneko K., ZSM-5 having uniform mesopore channels. *J. Am. Chem. Soc.* **125** (2003) pp. 6044-6045.
18. Tao Y., Kanoh H. and Kaneko K., Comparative Study on Pore Structures of Mesoporous ZSM-5 from Resorcinol-formaldehyde Aerogel and Carbon Aerogel Templating. *J. Phys. Chem. B.* **109** (2005) pp. 194-199.
19. Tao Y., Kanoh H. and Kaneko K., Synthesis of Mesoporous Zeolite A by Resorcinol-Formaldehyde Aerogel Templating. *Langmuir* **21** (2005) pp. 504-507.
20. Hanzawa Y. and Kaneko K., Lack of predominant adsorption of water vapor on carbon mesopores. *Langmuir* **13** (1997) pp. 5802-5804.
21. Miyamoto J., Kanoh H. and Kaneko K., The Addition of Mesoporosity to Activated Carbon Fibers by a Simple Reactivation Process. *Carbon* **43** (2005) pp. 855-857.
22. Lei S., Miyamoto J., Kanoh H. and Kaneko K., Enhancement of the methylene blue adsorption rate for ultramicroporous carbon fibers by the addition of mesopores. *Carbon* in press.
23. Kondo A., Noguchi H., Carlucci L., Mercandelli P., Procerpio D. M., Gianfranco C., Kajiro H., Kanoh H. and Kaneko K., Structural characterization of two dimensional metal-organic frameworks exhibiting an explicit adsorption jump. *J. Am. Chem. Soc.* in preparation.
24. Utsumi S., Miyawaki J., Tanaka H., Hattori Y., Itoi T., Ichikuni N., Kanoh H., Yudasaka M., Iijima S. and Kaneko K., Opening mechanism of internal nanoporosity of single wall carbon nanohorn. *J. Phys. Chem. B* **109** (2005) pp. 14319-14324.
25. Tanaka H., Kanoh H., Yudasaka M., Iijima S. and Kaneko K., Quantum Effects on Hydrogen Isotope Adsorption on Single-Wall Carbon Nanohorns *J. Am. Chem. Soc.* **127** (2005) pp. 7511-7516.
26. Bekyarova E., Murata K., Yudasaka M., Katsuya D., Iijima S., Tanaka H., Kanoh H. and Kaneko K., Single-wall nanostructured carbon for methane storage. *J. Phys. Chem. B* **107** (2003) pp. 4681-4684.
27. Murata K., Hashimoto A., Yudasaka M., Kasuya D., Kaneko K. and Iijima S., The use of charge transfer to enhance the methane-storage capacity of single wall carbon nanostructured carbon. *Adv. Mater.* **16** (2004) pp. 1520-1522.
28. Yang C.-Min, Noguchi H., Yudasaka M., Hashimoto A., Iijima S.and Kaneko K., Highly Ultramicroporosity-Donated Single-Wall Carbon Nanohorn Assemblies. *Adv. Mater.* **17** (2005) pp. 866-870.
29. Utsumi S., Urita K., Kanoh H., Yudasaka Y., Suenaga K., Iijima S. and Kaneko K., Preparing a magnetically responsive single-wall carbon

nanohorn colloid by anchoring magnetite nanoparticles. *J. Phys. Chem. B* **110** (2006) pp. 165-7170.

30. Urita K., Seki S., Utsumi S., Noguchi D., Kanoh H., Tanaka H., Ochiai Y., Aoki N., Yudasaka M., Iijima S. and Kaneko K., Effects of gas adsorption on the electrical conductivity of single wall carbon nanohorn. *Nano. Lett.* In press.
31. Ohba T., Kanoh H. and Kaneko K., Affinity transformation from hydrophilicity to hydrophobicity of water molecules on the basis of adsorption of water in graphitic nanopores. *J. Am. Chem. Soc.* **126** (2004) pp. 1560-1562.
32. Ohba T., Kanoh H. and Kaneko K., Structures and Stability of Water Nanoclusters in Hydrophobic Nanospaces. *Nano Lett.* **5** (2005) pp. 227-230.

EXPERIMENTAL METHODS FOR SINGLE AND MULTI-COMPONENT GAS ADSORPTION EQUILIBRIA

J. U. KELLER, N. IOSSIFOVA, W. ZIMMERMANN
Inst. Fluid- and Thermodynamics
University of Siegen, 57068 Siegen, Germany
E-mail: keller@ift.maschinenbau.uni-siegen.de

F. DREISBACH
Rubotherm Präzisionsmesstechnk GmbH,
Universitätsstr. 142, 44799 Bochum, Germany

R. STAUDT
Center of Non-Classical Chemistry,
Permoser Str. 15, 04318 Leipzig, Germany

An overview is given of classical and new experimental methods available today to measure adsorption equilibria of pure gases and gas mixtures on porous sorbent materials. These methods are: Volumetry / Manometry, Gravimetry / Densimetry, Oscillometry, Calorimetry, Impedance Spectroscopy and combinations thereof. The physical principles, advantages and disadvantages of these methods will be presented and discussed in brief [1]. Experimental data of Gibbs excess and / or absolute masses adsorbed will be presented. Recommendations are given for choosing the appropriate method if the purpose of measurements and requirements of accuracy and precision for either scientific or industrial needs are specified.

Introduction

Gas-solid equilibria data describe the amount of gas adsorbed on the (external and internal) surface of a given amount of a porous material at given pressure, concentration and temperature of the gas phase. These data are needed for

a) characterization of the porous solid used, i. e. the so-called sorbent, and
b) for design and evaluation of laboratory and industrialized gas adsorption processes used for separation and purification of gas mixtures or gases contaminated with environmentally hazardous components like FClHCs etc. [1].

The possibility for separating components of a gas mixture is due to the fact that interactions of molecules in the adsorbed phase are normally different from those in the bulk gas phase.

Equilibria data of pure or mixed gases on porous solids even today cannot be calculated from first principles, except in highly idealized systems which only have restricted relevance for technical processes [2]. Hence, they still have to be determined experimentally, i. e. by measurements which however for mixture gases often are laborious and cumbersome.

In this article a short overview is given of the measurement methods for adsorption equilibria of pure and mixed gases most often used today. After presenting the traditional volumetric and gravimetric method, modern combinations of it, namely the densimetric-volumetric and the densimetric-gravimetric method to measure binary coadsorption equilibria are presented in brief (Section 2).

In Sections 3 and 4 we will outline more sophisticated methods namely oscillometry for handling sorption equilibria in swelling sorbent materials like polymers and adsorption calorimetry for determining the heat of adsorption which is set free upon adsorption of a gas but needed for desorption of the adsorbed molecules form the sorbent material. Finally in Section 5 we will mention in brief impedance measurements in gas adsorption systems which still have potential to improve control of adsorption reactors on a commercial / industrial scale. Also hints are given for choosing a measurement method if the purpose of the measurements and requirements for the accuracy of data are given.

Measurement Methods

Equilibria states of pure or mixed gases adsorbed on the (external and internal) surface of porous materials like activated carbons or zeolites can be measured by using any of the basic physical properties of matter like its extensivity in space, gravity, inertia or molecular structure. An overview of these properties and resulting measurement methods is given in Table 1 below. Also, possibilities for combinations of these methods to measure gas mixture or so-called coadsorption equilibria are indicated.

In columns one and two the names of the various methods and their underlying physical properties of matter are given. In the upper right portion of the table (+) indicates availability and feasibility of the respective combination of methods. The symbol (0) means that this combination of measurement methods gives information on adsorption equilibria states of pure gases, but is not recommended for gas mixture measurements. The numbers in the lower left portion of the table indicate the number of adsorptive components for which the respective combination of the measurement methods can be applied.

Table 1. Measurement methods for adsorption equilibria of pure gases and gas mixtures on porous solids [1]. Explanations of the various symbols are given in the text of this article.

Method	Material Physics	V	G	O	SP	CH	D	C
Volumetry (V)	Extensivity		++	+	0	++	++	0
Gravimetry (G)	Gravity	2		+	0	+	+	0
Oscillometry (O)	Inertia	1, V	1, V		0	0	0	0
Spectroscopy (SP)	Electric Charges	1	1					
Chromatography (CH)	Molecules	N	N	(N)				
Densimetry (D)	Extensivity	2	2	1, V				
Calorimetry (C)	Thermal Inertia	1	1	1				

The most simple and still fairly reliable method to measure multi-component gas adsorption equilibria is the volumetric-chromatographic method. The basic installation for this method is sketched in Figure 1. It basically consists of a gas storage vessel of volume (V_{SV}) and an adsorption chamber of volume (V_{AC}) filled with adsorbent of mass (m^s) and provided with proper tubing and valves to allow gas circulation and evacuation. The gas (mixture) is first prepared in the storage vessel and then expanded to the adsorption vessel where it is partly adsorbed in the sorbent material.

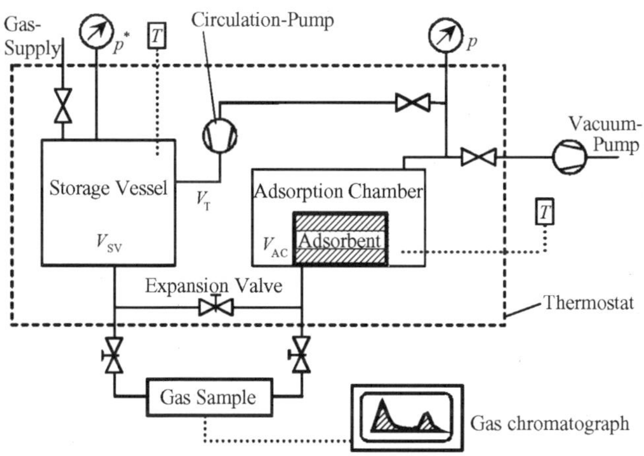

Figure 1. Experimental setup for volumetric-chromatographic measurements of multicomponent gas adsorption equilibria.

From the mass balances of all components and chromatographic measurements of all gas concentrations (w_i) in a gas chromatograph (GC) after equilibration the mass (m_i) of component (i = 1…N) adsorbed on (m^s) can be calculated as

$$m_i = (\rho_i^* - \rho_i^f)V_{SV} - (V_{AC} - V^s)\rho_i^f \tag{1}$$

$$\rho_i^f = w_i\rho^f(T, p, w_1...w_N), \qquad i = 1...N \tag{2}$$

Here $\left(\rho_i^*\right)$ is the partial density of component (i) initially realized in the storage vessel prior to adsorption and (T, p) indicate temperature and pressure in the adsorption vessel. V^s is the volume of the sorbent material, a quantity which can be approximated by its so-called He-volume [1].

In Figure 2 as an example coadsorption equilibria data of a ternary gas mixture ($CH_4 : CO_2 : N_2 = 48 : 8 : 44$ % mol) on activated carbon ACR1 (Norit) at 298 K for gas pressures up to 6 MPa are shown. This lines are correlation curves based on the 2-sites generalized Langmuir adsorption isotherm [1, 2]. Increasing deviations between measured and correlated data at increasing pressures should be observed.

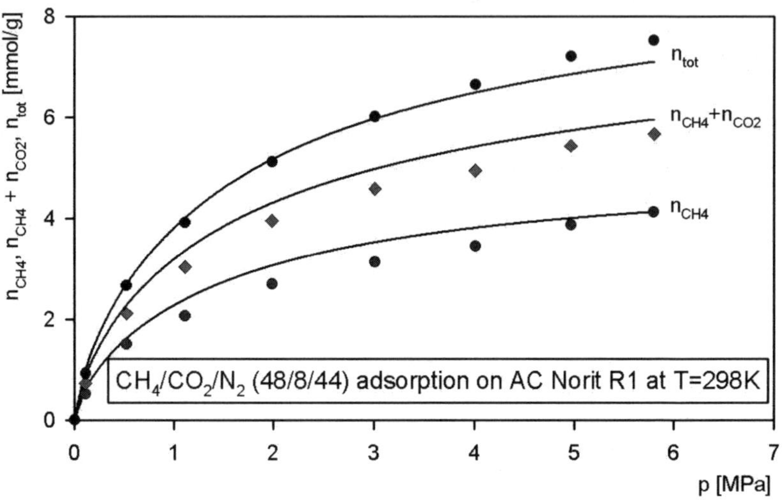

Figure 2. Adsorption equilibria of a ternary gas mixture ($CH_4 : CO_2 : N_2 = 48 : 8 : 44$ %mol) on ACR1 at 298 K.

In Figure 3 an experimental installation for volumetric flow measurements of adsorption equilibria of gas-solid-biocatalytic systems is given. The carrier gas flow (N_2 etc.) is augmented with substrate(s) like methanol (CH_3CH_2OH), glucose etc. and sent via a mixing chamber to the bioreactor(s) where the substrate is converted to product(s) by appropriate enzymes or bacteria. The product, for example acetic aldehyde (CH_3COH) and hydrogen (H_2) is released to the carrier gas and after concentration measurements in a GC easily separated from the carrier gas and remaining substrate by distillation etc. [3].

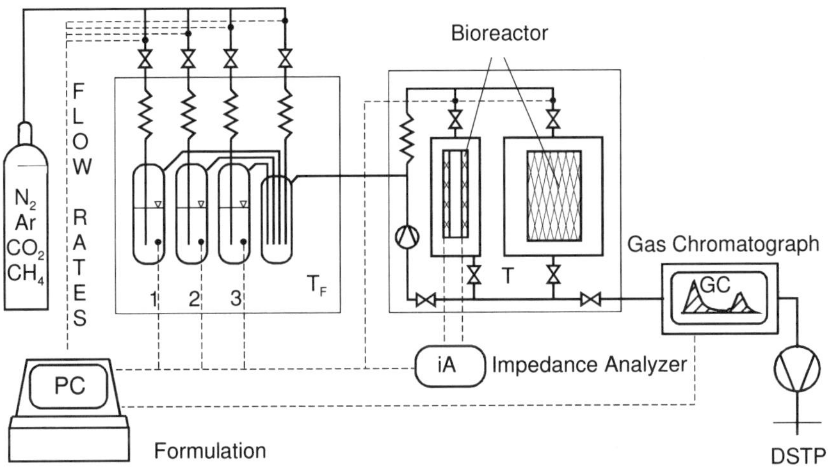

Figure 3. Volumetric-chromatographic analysis of gas-solid biocatalytic conversions as for example ethanol oxidation by dehydration: $CH_3CH_2OH \rightarrow CH_3COH + H_2$ enzyme from pichia pastoris [3, 4].

Main advantages of volumetric measurements are simplicity of installation and experimental procedure. Disadvantages are adsorption of the sorptive gas on the walls of tubes and vessels of the apparatus and uncertainty on whether or not equilibrium inside the adsorption vessel has been realized as this may take only seconds but sometimes many hours or even days.

In gravimetric-chromatographic measurements, i. e. by weighing the sorbent material sample, the approach to equilibrium, i. e. kinetics of the adsorption (and also desorption) process can be monitored. A schematic diagram of an installation for such measurements is given in Figure 4. It includes on its left side a magnetic suspension balance (Rubotherm GmbH, Bochum, Germany) allowing measurements with corrosive gases (H_2S, SO_2, etc.) [5]. The masses of an N-component adsorbate (m_i, i = 1…N) can be calculated from weighing data of the sorbent sample (Ω) and concentrations of the sorptive gas (w_i) after equilibrium has been established if those of the supply gas prior to adsorption $\left(w_i^*\right)$ are known:

$$m_i = w_i\left(\frac{p(V_{AC}-V^S)}{RTZ}M^f + \Omega\right) - w_i^f \frac{p(V_{AC}-V^S)}{RTZ}M^f, \qquad i = 1...N, \tag{3}$$

$$Z = Z(p,T,w_1...w_N), \qquad (M^f)^{-1} = \sum_i^N (w_i/M_i).$$

Here (Z) and (M^f) present the compressibility and the molar mass of the real gas adsorptive mixture respectively. For details refer to Ref. [1].

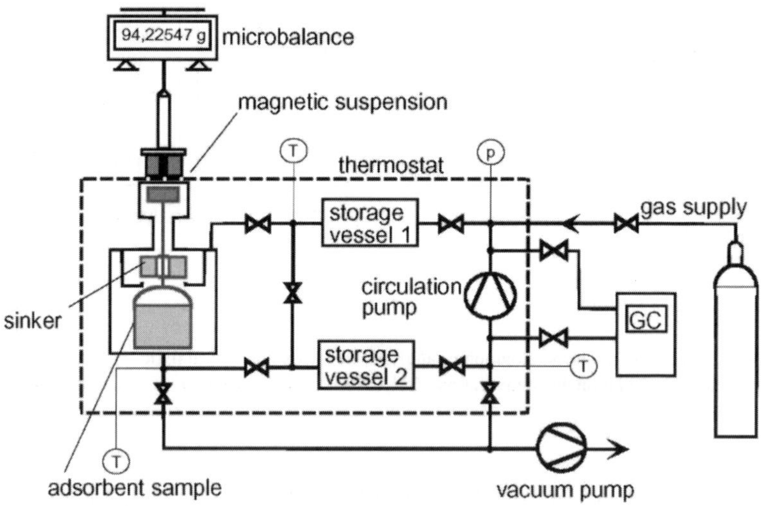

Figure 4. Schematic diagram of a gravimetric-chromatographic installation with a magnetic suspension balance for coadsorption measurements.

For binary coadsorption equilibria with non-isomeric gas components ($M_1 \neq M_2$) gravimetric-chromatographic measurements are not needed. Instead densimetric-volumetric measurements are recommended [6]. The measurement procedure can be grasped from the experimental scheme sketched in Figure 5 below. Basically, a gas expansion experiment is combined with a density measurement of the equilibrium sorptive gas mixture by the buoyancy of a sinker coupled to a magnetic suspension balance.

The masses of a binary gas mixture adsorbed on a sorbent material can be determined from combined pressure (p) and gas density (ρ^f) measurements. The resulting formulae are

$$m_i = m_i^* - \frac{M_i}{M_i - M_{i+1}}\left(\rho^f - \frac{pM_{i+1}}{RTZ(p,T,w_i)}\right)\left(V^* - V_{He}^s\right) \tag{4}$$

$$V^* = V_{SV} + V_{AC}, \quad i = 1,2(\mathrm{mod}\,2), \qquad M_i \neq M_{i+1}$$

Here again (Z) indicates the real gas compressibility of the adsorptive and (V_{He}^S) is the helium approximation of the sorbent's volume [1].

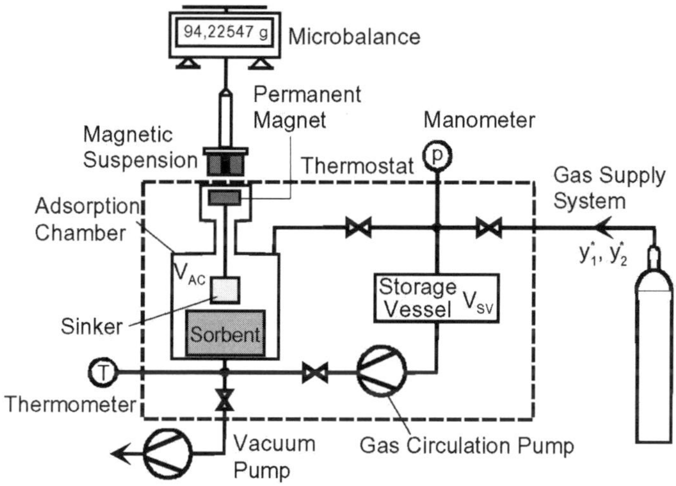

Figure 5. Densimetric-volumetric measurements of a binary coadsorption equilibria of premixed gases with molar concentrations (y_1^*, y_2^*).

Finally we would like to mention that binary coadsorption equilibria of non-isomeric gas components also can be measured without gas phase analysis by volumetric-gravimetric or gravimetric-densimetric, i. e. combined weighing and density measurements. Both procedures can be realized in an installation similar to that shown in Figure 4. Details are given in [1, Chapts 3, 4].

Adsorption equilibria measurement methods in swelling adsorbents

Polymers and other sorbent materials may change during ad- and desorption processes of gases not only their mass but also the volume, i. e. they swell or shrink during the sorption process. For such materials sorption equilibria can neither determined by volumetric or gravimetric experiments alone, but need additional measurements leading to two physically independent equations allowing to calculate both the mass and the volume of the resulting sorbent / sorbate system. One possibility for such measurements is given by slow rotational oscillations allowing to determine the (inert) mass of the sorbent / sorbate. Hence by weighing the sample its volume can be calculated from the buoyancy term of this measurement. A sketch of such a pendulum and a snapshot of a laboratory instrument are shown in Figure 6. An example of measured data is given in Figure 7 referring to sorption of carbon dioxide (CO_2) in polycarbonate (Bayer AG) at 293 K for pressures up to 6 MPa. Details of measurements and background theory are given in [1, Chap. 5].

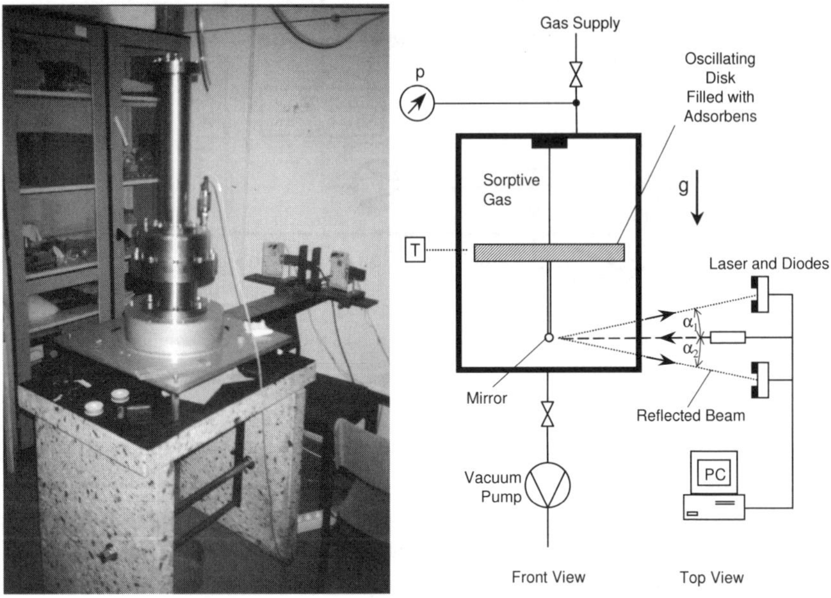

Figure 6. Rotational pendulum for measurements of gas adsorption equilibria by observing slow damped oscillations of a sorbent / sorbate system [1].

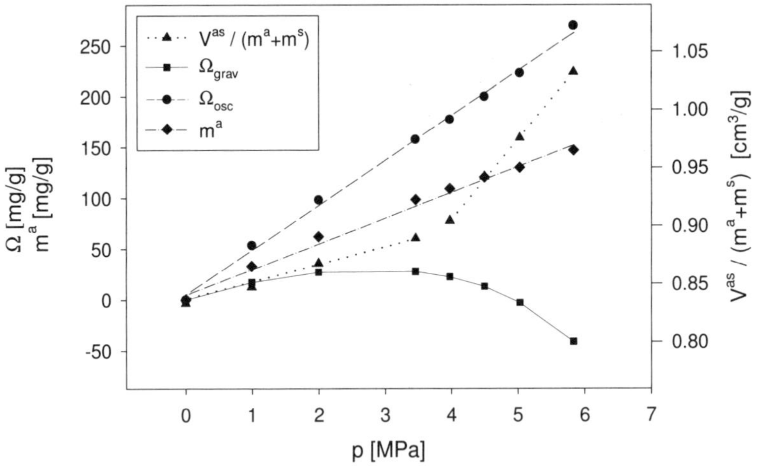

Figure 7. Change of mass (m^a) and specific volume ($V^{as}/(m^a+m^s)$) of a polycarbonate (Bayer AG) during sorption of subcritical CO_2 at $T = 293$ K $< T_{c\ co2} = 303{,}6$ K. The bend in the volume correlating line (▲) may indicate the glass transition point of the polycarbonate.

Adsorption Calorimetry

Adsorption processes of gases on porous solids are normally exothermic, the molar heat of adsorption being in the range (20 – 80) kJ/mol. Higher values indicate transition of reversible physisorption to irreversible chemisorption processes. If the heat of adsorption of a single molecule is known from molecular model calculation, the amount of gas adsorbed can be calculated from (integral) heat of adsorption measurements by dividing its numerical value by the molecular heat of adsorption [7]. A very effective instrument for heat of adsorption measurements is the so-called sensor gas calorimeter shown in Figure 8 below [8, 9]. Instead of usingthermocouples, it has a gas jacket surrounding the adsorption cell. A heat flow produced inside the cell will penetrate the sensor gas and thus increase both its temperature and its pressure. The time integral of the pressure signal is proportional to the total amount of heat released from the sorbent sample upon adsorption of gas. Figure 9 shows an example of such measurements referring to the adsorption of n-butane on activated carbon BAX1500 at 298 K, [8]. It should be noted that the SGC simultaneously allows measurements of heats of adsorption and also of the amount of gas adsorbed by a volumetric / manometric procedure.

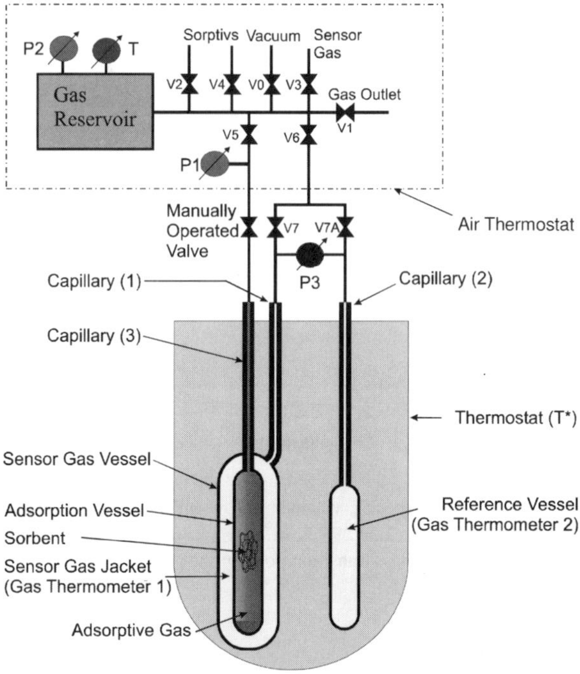

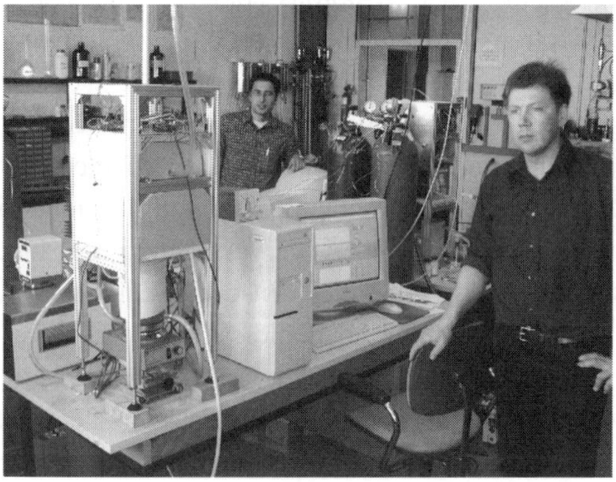

Figure 8. Schematic diagram of a sensor gas calorimeter (SGC) allowing simultaneous measurements of the heat and the mass of a gas adsorbed on a sorbent sample [8]. On the right hand side a laboratory scaled instrument and auxiliary equipment (stirrer, gas supply system, PC etc.) is shown as is used at IFT, University of Siegen, since 2003.

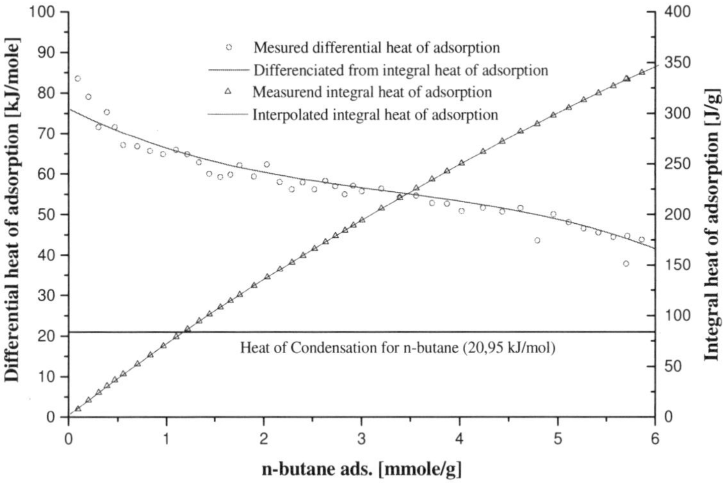

Figure 9. Differential and integral heat of adsorption of n-butane gas on activated carbon BAX 1500 at 298 K measured with a sensor gas calorimeter [8].

Dielectric permittivity measurements

The ratio between the dielectric displacement vector (**D**) and that of the electric field strength (**E**) is called the dielectric permittivity (ε) of a material : $\varepsilon \equiv \varepsilon_r \varepsilon_0 = (\mathbf{D}/\mathbf{E})$. Here $\varepsilon_0 = 8.8542 \cdot 10^{-12}$ As/Vm indicates the permittivity of vacuum and (ε_r) is the so-called relative permittivity of the material. As ε_r depends on magnitude and spatial arrangement of all electric charges included in a material, it changes if gas is either adsorbed or desorbed in the material. Indeed, the absolute value of (ε_r) can be considered as measure, i. e. a linear function of number of gas molecules adsorbed in the material [1, Chap. 6, 9, 10].

An example for permittivity measurements is given in Figure 11. It shows the real part of the complex capacity (C = C(f, T)) as a function of the frequency (f) of the (weak) oscillating electric field applied to the capacitor for the zeolite DAY-carbon dioxide (CO_2) system at 298 K. The lowest line refers to vacuum, the upper line to the maximum gas pressure of 1.9924 MPa. Note that all curves are shifted monotonously to higher capacity values as the pressure of the gas and thus the amount of CO_2 adsorbed increases.

Impedance measurements inside an adsorption reactor can be used as local manometers or as indication of local accumulation of (preferably) polar sorbate components as for example carbon monoxide in activated carbon adsorbers. This component provides an early warning for "hot spots" inside the reactor and often occurs prior to inflammation and burning. An example for this type of measurements is shown in Figure 12. Here combined pressure (p) and impedance/capacity data are shown as function of time which have been taken inside an industrial sized adsorption reactor designed for air separation processes [1, Chap. 6]

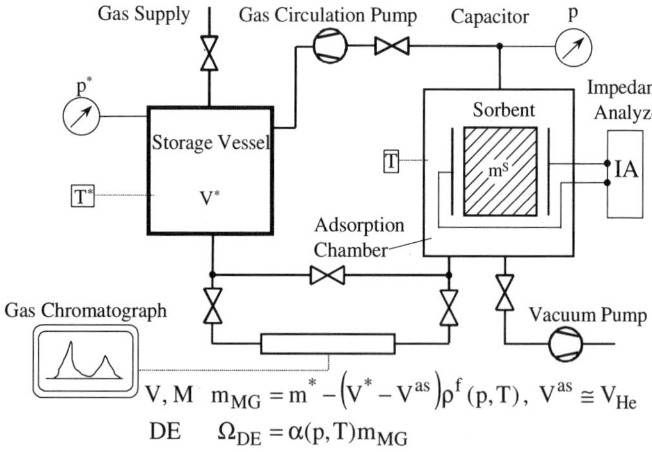

Figure 10. Experimental setup for simultaneous volumetric-dielectric measurements to determine the amount of gas adsorbed and the dielectric permittivity of a sorbent / sorbate system.

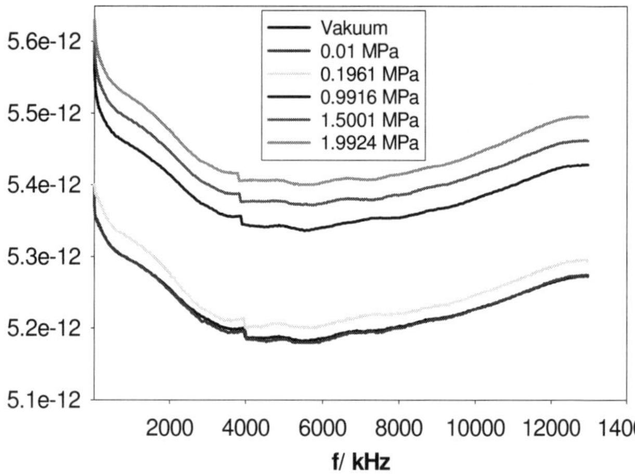

Figure 11. Dielectric impedance or capacity measurements of carbon dioxide (CO_2) adsorbed on zeolite DAY (Degussa) at 298 K.

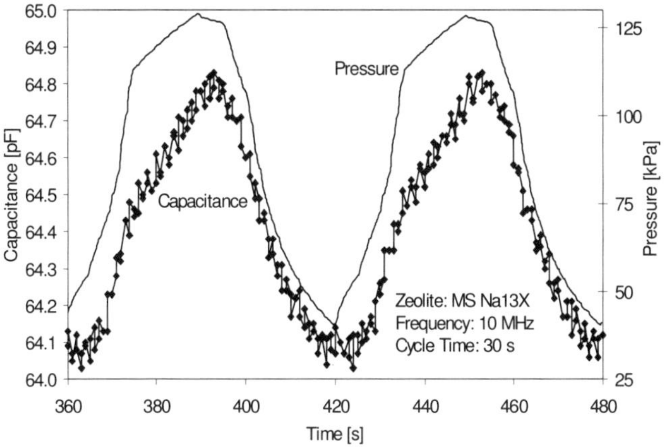

Figure 12. Combined pressure (p) and dielectric (ε_r) measurements of a periodic ad- and desorption process of nitrogen (N_2) on molecular sieve MSNa13X (UOP) at 293 K taken inside an industrial sized adsorption column (PSA).

Conclusions

Today there are several experimental methods available to measure pure gas and gas mixture adsorption equilibria on porous rigid or swelling sorbent materials. All these methods have their specific advantages and disadvantages [1]. Choice of any of them depends mainly on the purpose of measurement and/or accuracy and reliability of data needed. For quick measurements of restricted accuracy gas expansion experiments or volumetric measurements are recommended. If high accuracy data are needed, weighing procedures, i. e. gravimetry should be used

Table 2. Measurement methods for gas adsorption equilibria as related to purpose of measurement and/or quality of data needed, cp. also Table 1.

Pure Gas Method	**Purpose**
Volumetry/Manometry	Characterization of porous solids
Gravimetry	Equilibria, Kinetics, Gas Density, Process
Oscillometry	Cont.
Dielectric Permittivity	Swelling Material
	Industrial Process Control
Gas Mixtures (N=2)	
Volumetric-Densimetric M.	Equilibria, Process Control
(2-sites Magnetic Balance)	
Gas Mixtures (N>2)	Process Design
Volumetric/Gas Phase Analysis	

as it on principle allows to monitor the approach to equilibrium of the gas-solid adsorption system. A brief overview of main purposes of measurements and recommended experimental methods is given in Table 2. For detailed discussion of all the experimental methods the reader kindly may refer to the literature cited, esp. Ref. [1].

Acknowledgements

The authors are grateful to many colleagues from all over the world who by discussions at international meetings (VMT, FoA, COPS, AIChE, PBCAST etc.) have contributed directly and indirectly to the development and evaluation of the measurement methods of gas adsorption equilibria presented in this article.

References

1. Keller J. U. and Staudt R., Gas Adsorption Equilibria, Experimental Methods and Adsorption Isotherms, p. 421, *Springer, New York, USA, ISBN 0-387-23597-3.*
2. Iossifova N., Untersuchungen von Gemischgleichgewichten bei adsorptiven Gastrenn- und Reinigungsverfahren, *Fortschrittberichte VDI,* Reihe 3, Verfahrenstechnik, VDI-Verlag, Düsseldorf, in preparation (2006)
3. Laware S., Legoy M.-D. and Graber M., Solid / gas bioreactors: powerful tools for fundamental research and efficient technology for industrial applications. *Green Chemistry* **Vol. 6** (2004) p. 445.
4. Bousquet-Dubouch M.-P. et al., Alcoholysis catalyzed by Candida antarctica lipase B in a gas / solid system obeys a Ping Pong Bi Bi mechanism ..., *Biochimica et Biophysica Acta,* **1550** (2001), 90–99.
5. Rubotherm Präzisionsmesstechnik GmbH Suspension Balances, International Application Notes, available from *Robotherm GmbH, Universitätsstr. 142, D-44799 Bochum, Germany,* www.rubotherm.de, 2001.
6. Keller J. U., Iossifova N. and Zimmermann W., Volumetric – Densimetric Measurements of the Adsorption Equilibria of Binary Gas Mixtures, *Adsorption Science & Technology*, 23 (No. 9) (2005) p. 285–702.
7. Guillot A., Stoeckli F. and Banguil Y., The Microporosity of activated carbon fibre KF1500 assessed by combined CO_2 adsorption and calorimetry, *Adsorption Science and Technology,* **18** (2000) p. 1–14.
8. Zimmermann W. and Keller J. U., A new calorimeter for simultaneous measurements of isotherms and heats of adsorption, *Thermochimica Acta* **405** (2003) p. 31–41.

9. Jackson J. D., Classical Electrodynamics, *J. Wiley & Sons, New York.*, **2nd Ed.,** (1975).
10. Frohlich H., Theory of Dielectric Constants and Dielectric Loss, *Oxford Science Publ., Oxford, UK,* Reprint 1986.

EXPERIMENTAL DETERMINATION OF HEAT EFFECTS THAT ACCOMPANY SORPTION EQUILIBRIUM PROCESSES

MARTIN BÜLOW

Am Rökerberg 22, D-18347 Ostseebad Dierhagen, Germany

Development of the sorption-isosteric method with minimum dead volume for a direct measurement of sorption heats in gas-nanoporous-sorbent systems is reviewed. Advantages and limitations of the technique are assessed and illustrated by concentration dependences of the isosteric sorption heat for various systems, several of which are discussed in the light of molecular simulation. The technique is useful and effective in obtaining highly accurate sorption-thermodynamic data for single gases and gas mixtures by nanoporous materials, *e.g.*, zeolites. These sorption-energetic properties are accessible as functions of sorption-phase concentration up to saturation values. They also serve for calculation of sorption isostherms for single gases and their mixtures over wide ranges of temperature and pressure - irrespective of phase transitions that may occur in the system.

1. Introduction

The author dedicates this paper to the memory of Professor *Lovat V.C. Rees*, Edinburgh, Scotland. He had been a personal friend of Professor *Rees* for some 25 years, and it is with greatest sadness to hear of his death on May 1, 2006.

Gas-solid sorption-thermodynamic data such as enthalpy, standard entropy, standard *Gibbs* free sorption energy and heat capacities of sorption systems, are important parameters in designing and modeling industrial separation and purification processes. Although having been an important research topic for decades [1], their correct determination still represents a challenge even nowadays, due to an ongoing intense development of novel sorbents and processes, in particular for sorption systems with relatively weak sorptioninteraction forces, or if individual sorbing components of a fluid mixture have similar sorption properties. On the other hand, during recent years, significant progress has been made in the field of simulation of sorption processes by *Monte Carlo* and Molecular Dynamics methods, first of all due to basic methodical reasons and computational hardware development. Much of their further success rests, however, on an availability of highaccuracy experimental data, in particular for the energetics of sorption phenomena, and on a close collaboration between groups that work theoretically and experimentally. The four most-widely used experimental methods to investigate sorption energetic properties comprise the following: differentiation of sorption isotherms at constant sorption-phase concentration, calorimetric methods, which can be executed under various conditions, direct measurement of sorption isosteres, and adsorption gas-chromatographic method [1-5]. Each of these methods ought to be developed further with regard to both its specific technical substance and in conjunction with other methods, which allows for their mutual control and stimulation.

This paper deals with the principles, advantages and limitations of measurement of sorption equilibria under isosteric conditions. It further assesses the sorption-isosteric method (SIM) as an effective tool for providing complete sets of sorption-thermodynamic functions, *viz.*, enthalpy, standard entropy and standard *Gibbs* free energy of sorption, for nanoporous solids, *i.e.*, micro- and mesoporous ones, as functions of sorption-phase concentration, *n*, over its entire range, and to approach such data for mixtures. The usefulness of SIM is exemplified by sorption systems that comprise atmospheric gases on zeolites and carbon dioxide, CO_2, on carbonaceous sorbents, as well as several of their mixtures.

2. History of the Sorption-isosteric Method

The basic idea of direct measurement of sorption isosteres for microporous sorption systems was first expressed by *Serpinsky* in 1967 [6] and published by *Bering et al.* in 1969 [7]. Fundamental thermodynamic features related to sorption isosteres and their direct measurement were discussed frequently by *Bering*, *Serpinsky*, *Fomkin et al.*, *e.g.*, in [8-11]. The first direct measurement of single-component sorption isosteres was carried out by the *Schirmer* school for n-paraffin compounds on FAU- and LTA-type zeolites, reported in 1969 and published in 1971 [12]. Extended basic research performed by that school, specifically for hydrocarbon-zeolite systems, utilized SIM in close connection with other techniques, *e.g.*, calorimetry, and theoretical methods such as *Monte Carlo* and statistical thermodynamics [13-17].

A first SIM investigation of sorption-thermodynamic functions for binary [18-20] and ternary [21] mixtures of gases on microporous solids was presented by the *Bülow* group, in the nineteen eighties and 1994, respectively; *Bülow* also introduced this technique to the *Rees* group at the ICSTM London [22]. Since 1989, the latter group published a series of papers, particularly on sorption equilibria for binary mixtures [23-26]. Thermodynamic analyses of the isosteric principle and of isosteric heats of multi-component sorption were performed by *Sircar* [27] and *Karavias* and *Myers* [28], respectively. Unfortunately, basic advantages of SIM were overlooked in [27] as they were in [2,3].

Since 1993, SIM had been improved significantly by using advanced automated technologies such as computerized controls for data acquisition and analysis to obtain highquality single-component and mixture-sorption thermodynamic data [29 and 30]. Related reports were published by *Bülow* and *Shen* in another number of articles [31-37], partly in collaboration with other laboratories [36-38]. A modern SIM version and its great utility were portrayed in [30]. The method to predict total mixture-sorption thermodynamic functions and extensive experimental information of that paper were republished in [38]. Utilization of SIM for an advanced characterization of sorption properties of nanoporous materials has contributed successfully to the development of several BOC proprietary sorbents for gas separation and purification, specifically for oxygen VPSA processes (Li,RE-LSX zeolite [39], RE: Rare Earth metal cations), and the removal of CO_2 from air streams up-front cryogenic air separation (NaLSX zeolite [40]).

3. Basic Principle of the Sorption-isosteric Method

3.1. *Theoretical*

The basic principle of SIM follows from a fundamental phenomenological experience that stems from basic research executed in the area of physical sorption over many decades, *viz*., sorption isosteres may presumptively be considered as straight lines at constant sorption-phase composition, n = const., in *Clausius-Clapeyron* plots, ln p *vs*. 1 / T. In accordance with [41-43], this finding allows to calculate the differential molar sorption heat, Q, as difference between the molar enthalpy of the gas phase, Hg, and the partial molar enthalpy of the sorbed substance, $\overline{H}_n$:

$$Q=\left(1-\frac{v_n}{v_g}\right)RZ\left(\frac{\partial \ln p}{\partial 1/T}\right)_n = H_s-\overline{H}_n=\Delta H=-q_{st} \qquad (1)$$

where p and T denote, respectively, gas-phase equilibrium pressure of sorbing species and absolute temperature; R stands for the universal gas constant; Z is the compressibility coefficient, $Z = pv_g / RT$ (Z = 1 for an ideal gas phase and $Z \neq 1$ for a real gas phase); v_n and v_g denote the partial molar volume of sorbing species in the sorption phase, $v_n = (\partial v/\partial n)_{p,T,no}$ (n_o denotes volume and mass of sorbent), and their molar volume in the gas phase, respectively. If the sorbent remains inert during the sorption process, *i.e.*, $v_n = 0$, the value of the slope of plot, ln *p vs.* 1/*T*, apparently presumed to be an "isostere", multiplied by R at $n =$ const., is also known as the isosteric heat of sorption, q_{st} (*cf.*, also ref. (10)). The quantity q_{st} differs from the differential heat of sorption, $-\Delta\overline{H}$, by the mechanical-work term $\mid RT \mid$: $q_{st} = -\Delta H = \Delta\overline{H} + RT$. During measurement of "apparent" sorption isosteres, one has to check very carefully whether or not the experimental curves, ln *p vs.* 1/*T*, were indeed straight lines within specifically formulated limits to variations allowed for the experimental measurables. In principle, the linearity of plot, ln *p vs.* 1/*T*, is an approximation, and it may or may not be valid for the following reasons:

(*i*) According to the *Kirchhoff Law*, the differential heat of sorption as any reaction enthalpy depends on temperature:

$$\left[\frac{\partial(\Delta H)}{\partial T}\right]_p = \Delta C_n(T) = C_n^{(ssyst)}(T) - \left[C_n^{(sorb)}(T) + C_n^{(sspec)}(T)\right] \tag{2}$$

$$\Delta C_n(T) \neq 0 \quad \Rightarrow \quad -\Delta H = f(T), \tag{3}$$

where $\Delta C_n(T)$, $C_n^{(ssyst)}(T)$, $C_n^{(sorb)}(T)$, and $C_n^{(sspec)}(T)$ are the specific heat-capacity change at $n =$ const., and the specific heat capacities for the overall sorption system, the sorbent and the sorbing species, respectively. This implies validity of $\Delta C_n(T) = 0$ if an isostere is linear, or, as for eq. (3), the isostere is not linear, *cf.*, [44], which makes it either an "apparent" one, or demonstrates existence of *T*-dependent sorption states, *cf.*, case (*ii*).

(*ii*) Phase sorption-phase transitions may occur, *cf.*, [11-17], which could lead to two straight branches of an isostere with particular (asymptotic) slopes that correspond to two specific isosteric sorption heats, $-\Delta H_i$, characteristic of the two sorption states. In analogy to an equilibrium reaction system [45], these transitions, *e.g.*, of the type "order ⇔ disorder", in particular "localization ⇔ delocalization", contribute to the overall change in the specific heat capacity, ΔC_n, at $n =$ const. of the sorption system as follows:

$$\Delta C_n = \left[\frac{\left[-\Delta(\Delta G^o{}_i)\right]^2}{RT^2}\right]_n \cdot \left(\frac{\exp\left[-\frac{\Delta(\Delta G^o{}_i)}{R}\left(\frac{1}{T_o}-\frac{1}{T}\right)\right]}{\left\{1+\exp\left[-\frac{\Delta(\Delta G^o{}_i)}{R}\left(\frac{1}{T_o}-\frac{1}{T}\right)\right]\right\}^2}\right)_n \tag{4}$$

where $\Delta(\Delta G^o{}_i)$ and T_o denote the difference in the changes of standard *Gibbs* free sorption energy between the two phase states and a "transition temperature", T_o, *i.e.*, at the "crosspoint" of the two asymptotic isostere branches, where the index i (= 1,2) refers to the two sorption states. Neglecting the entropy term, $\Delta(\Delta S^o{}_i)$, eq. 4 can be rewritten approximately in terms of an isosteric sorption-heat difference, $\Delta(\Delta H_i)$. Direct caloric measurement [46, 47] of dependences, $\Delta C_n^{(ssyst)}(T)$, have suggested to consider sorption-phase transitions in nanoporous solids rather like *Schottky*-type than λ-point anomalies [48]. Over the past decades, measurement of specific heat capacities of sorption systems has attracted little attention only, *cf.*, [5 (and quotes therein), 46-51] despite tremendous value of such information. A combination of the *Clausius-Clapeyron* equation with the *Kirchhoff Law* leads to the following general expression for a sorption isostere, *viz.*, at n = const.:

$$R\left[\frac{\partial \ln p}{\partial\left(\frac{1}{T}\right)}\right] = \Delta H_o + \int_0^{T_o} \Delta C_n dT + \Delta C_n T_o + \int_{T_o}^{T} \Delta C_n dT + \ldots \tag{5}$$

Neglecting the contradiction between relationships (3 and 5), on the one hand, and, on the other hand, the *Clausius-Clapeyron* equation in its simplified shape (eq. 6) (isosteres are found to be linear over very broad regions of T, p and n),

$$q_{st} = -\, R\left[\frac{\partial \ln p}{\partial\left(\frac{1}{T}\right)}\right]_n , \tag{6}$$

utilization of the latter one becomes justified, probably, as a result of a compensation effect due to the use of pressure instead of fugacity and neglecting the molar volume of the sorption phase ("condensed" phase) with regard to that of the gas phase [52]. (Sorption-phase transitions in nanoporous systems will be discussed in more detail by a paper in preparation [53]) (*iii*) An "apparent" sorption isostere may deviate from linearity due to in(de)creasing desorbed amount with *T* in(de)crease to an extent that is specific for a considered sorption system, over a given *T* range. A correction of such a de(ad)sorbed amount can be applied to a single-component sorption isostere based on considerations below. An analogous correction is practically impossible for the mixture-sorption case, because of exact knowledge needed for mixture-sorption isotherms, which is very difficult to obtain.

If an isostere is not linear due to non-negligible de(ad)sorption , $n \neq$ const., that results from *T* in(de)crease during an "isosteric" experiment, which would lead to an error in sorption-phase concentration by

$$\Delta n_s = n_s^{(1)} - n_s^{(2)} \tag{7}$$

where $n_s^{(2)}$is the dosed amount of species in the sorption phase, $n_s^{(1)}$ is the real sorbed amount in the sorption phase, and $\Delta n_s \approx p_1 V_d / RT_1$, where V_d is the "dead volume" of the SIM sorption cell, T_1 is the equilibrium temperature of the system, and $p_1 = f(n_s^{(1)})$ represents the equilibrium pressure. The pressure increment, Δp, caused by de(ad)sorption can be calculated and used to correct the equilibrium pressure measured under isosteric conditions,

$$\Delta p = \frac{\partial p}{\partial n_s} \Delta n_s \quad \Rightarrow \quad p_2\big|_{T_1} = p_1^{(meas.)} + \Delta p \,, \tag{8}$$

where $\frac{\partial p}{\partial n_s}$ represents the reciprocal slope of the sorption isotherm for T_1 at $(n_s^{(1)}, p1)$ measured independently, which reads as follows:

$$\Delta \ln p = \left.\frac{\partial \ln p}{\partial n_s}\right|_{\substack{T=T_1 \\ n_s = n_s^{(1)}}} * \frac{p_1 V_d}{RT_1} \,. \tag{9}$$

3.2. *Thermodynamic Description of Mixture Sorption*

"Surface free energy", $A\pi/n_s$, which - for a microporous material - can be determined from sorption isotherms as a complex quantity only, without splitting it into numerical values of specific surface area, *A*, spreading pressure, π, and number of moles of the sorbent, *n*s, and which should be considered as change of the chemical potential, Δμ, of the sorbent as a result of the sorption process [54], can be calculated directly from sorptionthermodynamic functions. For this purpose, these functions that characterize the sorption process, can be expressed by their polynomial fits with regard to sorption-phase concentration, n. The following equations are used:

$$A\pi = \mathrm{F}(n) = \int n\ \mathrm{d}(\Delta G) = \int n\ \mathrm{d}(\Delta H) - T\int n\ \mathrm{d}(\Delta S) \qquad (10)$$

$$\Delta H = \mathrm{A} + \mathrm{B}n + \mathrm{C}n^2 + \mathrm{D}n^3 + \ldots\ldots \qquad (11)$$

$$\Delta S = \mathrm{a} + \mathrm{b}n + \mathrm{c}n^2 + \mathrm{d}n^3 + \ldots\ldots \qquad (12)$$

$$A\pi = \left(\frac{\mathrm{B}}{2}n^2 + \frac{2\mathrm{C}}{3}n^3 + \frac{3\mathrm{D}}{4}n^4 + \ldots.\right) - T\left(\frac{\mathrm{b}}{2}n^2 + \frac{2\mathrm{c}}{3}n^3 + \frac{3\mathrm{d}}{4}n^4 + \ldots.\right). \qquad (13)$$

By combining eqs. (10-13) with the *Gibbs* function, eq. (14),

$$\Delta G^{o}(n) = \Delta H(n) - T\Delta S^{o}(n) \qquad (14)$$

- this approach being called “Adsorbate Solution Theory” (AST) to distinguish it from the “Ideal Adsorbed Solution Theory” (IAST) [55] -, one may predict total mixturesorption thermodynamic functions from those for single components, at constant changes of chemical potential of the sorbent and at constant temperature and sorption-phase composition:

$$\exp\left(\frac{\Delta G_m(n_m)}{\mathrm{R}T}\right) = \sum_i^N x_i\,\gamma_i\ \exp\left(\frac{\Delta G_i^o(n_i^o)}{\mathrm{R}T}\right), \qquad (15)$$

where the functions $\Delta G^{o}{}_{i}(n^{o}{}_{i})$ denote the concentration-dependent changes of singlecomponent *Gibbs* free sorption energy at the same value of “surface free energy”, as that of the binary mixture, *cf*., eq. (16):

$$(A\ )m_\pi = (A_\pi\ )^o{}_1 = (A_\pi\ )^o{}_2 \qquad (16)$$

From plots of “surface free energy”, A_π / n_s, v_s. sorption-phase concentration, n, at a given temperature, total mixture-sorption isotherms at constant sorption-phase composition, $x_i(s)$, can be calculated using the following formalism, where pm denotes the total pressure of the mixture at sorption equilibrium:

$$p_m = 760\sum_i^N x_i^{(s)}\gamma_i^{(s)} \exp\left(\frac{\Delta G_i^o}{\mathrm{R}T}\right) \qquad (17)$$

$$\left(\frac{1}{n_m}\right) = \left(\frac{1}{n_m}\right)^e + \sum_{i=1}^N\left(\frac{x_i^{(s)}}{n_i^o}\right). \qquad (18)$$

Activity coefficients, $\gamma_i(s)$, of component i in the sorption phase can also be calculated. For this purpose, the following fundamental thermodynamic expressions could be used:

$$(\Delta G)_m^{\ e} = (\pi A)_m \left(\frac{1}{n_m} \right)^e \tag{19}$$

$$\frac{(\Delta G)_m^{\ e}}{RT} = \sum_{i=1}^{N} x_i^{(s)} \ln \gamma_i^{(s)} \ . \tag{20}$$

Although this feature of mixture sorption will not be addressed in detail in this paper, its utility will be exemplified below.

3.3. *Methodical*

The experimental execution of SIM comprises a consecutive measurement of equilibrium pressure p as function of T at $n \cong$ const. in a closed system with co-existing gas-solid phases. It is executed experimentally with a minimum dead volume, to ensure presence of only a comparatively negligible faction of sorbing species in the gas phase, and, thus, to indeed maintain a (nearly) constant sorption-phase concentration, n, even if T being changed ("isosteric" refers correctly to the complex of co-existing sorption and fluid phases). It should be understood that the emerging deviations in parameter n, when T changes, are - as a rule - within the error margin of the determination of concentration n in cases of other experimental methods such as isotherm measurement or calorimetry.

Sorption-thermodynamic functions as dependences on concentration, n, *e.g.*, the isosteric molar sorption enthalpy, $\Delta H(n)$, the standard sorption entropy, $\Delta S°(n)$, and the standard *Gibbs* free sorption energy, $\Delta G°(n)$, can be calculated by basic formulas (21), (22) and (14), respectively,

$$\ln p = const. - \frac{q_{\text{isosteric}}(n)}{RT} \quad , \tag{21}$$

$$\Delta S°(n) = \frac{\Delta H(n)}{T} - R \ln\left(\frac{p}{p°} \right) , \tag{22}$$

by repeating those measurements for different values of n, the latter being controlled by volumetric dosing procedures. If needed, the isosteric heats can be used to calculate integral sorption heats over defined ranges of sorption-phase concentration. By shaping appropriately both experimental device and experimental execution of the method, the above-described inherent contradictions of the principal idea of "isosteric" measurements can be minimized sufficiently. A successfully utilized SIM version as outlined schematically in Figure 1 is characterized by the following main features:

(*i*) Minimum dead volume: minimum void volume and large amount of sorbent, *c.* (5 ~ 15) g; (*ii*) minimum gas-phase volume to sorption-phase volume ratio, $Vg/Vs < 5$; (*iii*) low p at equilibrium, (0.0133 ~ 13.337) kPa; (*iv*) small T increments, *c.* (2 – 5) K; (*v*) strongly controlled equilibration criteria for both T and p, and high accuracy of their measurement (feature neglected in related assessment of SIM [2]); (*vi*) equilibria can further be controlled by changing T in different directions at $n \cong$ const.; (*vii*) highaccuracy dosing procedure at entirely thermostated conditions; (*viii*) gas-phase circulation in the SIM sorption cell; (*ix*) sophisticated data-acquisition and evaluation software; (*x*) apparatus layout for measurements at cryogenic temperature; (*xi*) any violation of the isosteric condition due to experimental reasons, *i.e.*, de(ad)sorption, becomes directly visible (feature neglected in related analysis [27]); (*xii*) occurrence of phase transitions are monitored sensibly.

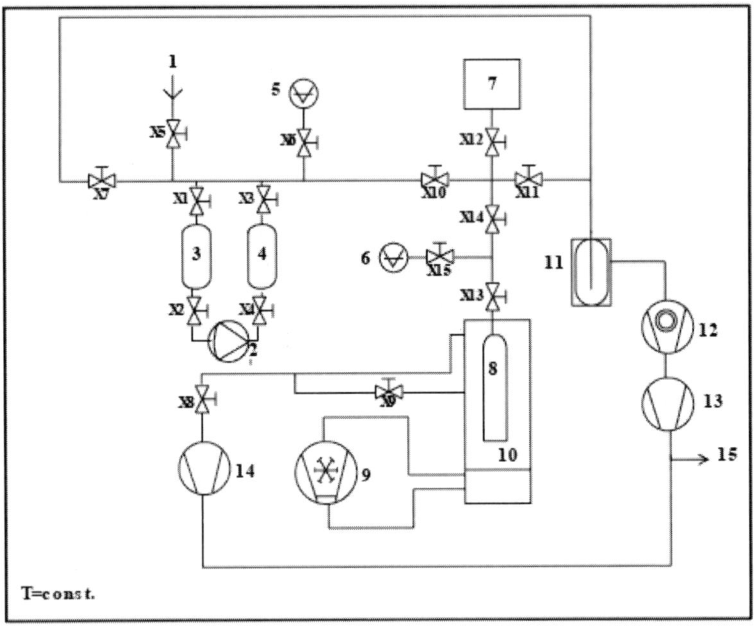

Figure 1. Scheme of SIM Apparatus. 1. Gas supply; 2. Circulating pump; 3&4. Gas cylinders; 5&6. Pressure sensors; 7. MS; 8. Sample holder; 9&10. Cryostat; 11-15, Vacuum systems.

The accuracy of SIM was proven, *inter alia*, by measuring the sublimation curve of CO_2 in the absence of sorbent [30]. The resulting changes of enthalpy, -25.26 kJ/mol, and entropy, -129.57 J/mol K, typical of CO_2 sublimation, agree with literature data [56] that amount to - 25.23 kJ/mol and -129.63 J/mol K, respectively. In terms of sublimation energy, the experimental accuracy is *ca.* ± 0.07 kJ/mol. Concerning isosteric sorption heats, *qst*, the experimental accuracy can be further increased by choosing sections of sorption isosteres with highest slope at given concentration. This approach is due to the experience that external influences on a sorption system lead to a decrease in isostere slope. The determination of highest slopes represents a special feature of the dataacquisition software utilized, in conjunction with a high-performance helium-cryostat system. Additional accuracy is gained in regions of cryogenic temperature because for a given constant temperature interval, ΔT, the interval, $\Delta(1/T)$, on the *abscissa* scale is spread out at low absolute temperature compared with that at high absolute temperature. This leads to a more accurate determination of the isostere slope measured at cryogenic temperature over the same temperature interval, ΔT. Altogether, this combination enables the current technique to minimize the experimental error of *qst* to *c.* ± 0.05 kJ/mol.

In case of multi-component mixtures, total isosteres can be measured at constant sorption-phase composition by changing, in successive steps, the total amount of gas mixture sorbed at constant mole fractions of sorption phase [31]. On the other hand, a point-bypoint measurement of partial pressures of a multi-component mixture sorbed leads to partial mixture-sorption isosteres that can be evaluated further by solution thermodynamics [57].

The *MSI* Cerius2 3.8 software package was used to study physical sorption of N_2 and O_2 on LiLSX zeolite as function of pressure of the sorbing species. Calculations are based on the application of a *Monte Carlo* simulation algorithm in the Grand Canonical Ensemble [58,59]. The interaction-potential parameters used in the forcefield expression of this investigation are published in [60], together with details of the simulation setup.

3.4. *Experimental Consistency Check of Isosteric Sorption Heats*

Following pioneering work of *Kiselev* who had been the first to compare sorption heats of identical systems measured by different techniques [1a], another experimental consistency check of 70-200 K, respectively. Temperatures for SIM heats of CO_2 on NaX are 155-310 K. Calorimetric heats were measured at 195 K for N_2 and O_2 on CaA and at 298 K for CO_2 on NaX pellets. The comparison is shown in Figure 2. For all systems compared, the SIM data is in

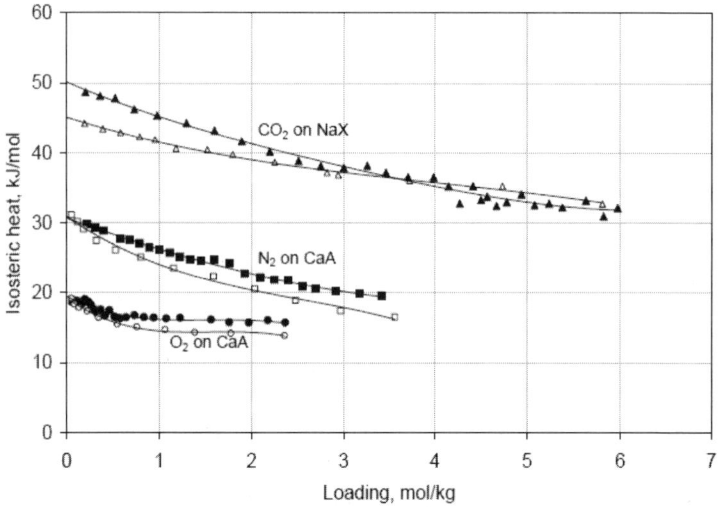

Figure 2. Comparison of heats of sorption for N_2 and O_2 on CaA and CO_2 on NaX zeolites measured with SIM (open symbols) and *Tian-Calvet* calorimetry (full symbols).

reasonable agreement with the calorimetric one, but the calorimetric heats are on average - by about 2 kJ/mol higher than the SIM heats.

4. Experimental Results and Discussion

4.1. *Sorption Heats of Atmospheric Gases on NaLSX Zeolite*

Sorption isosteres were measured for CO_2, N_2O, N_2 and O_2 on NaLSX pellets (13 wt.-% binder; Si/Al mole ratio of pellets: 1.28), coded as FAU-I (*cf.*, Table 2 in [30]), over wide ranges of p, T and n as seen from the sorption isosteric plots in Figures 3-6, respectively.

Compared to the sorption isosteres of the three other gases, those of N_2O at high sorption- phase concentration show specific shapes [61], which could be attributed to the existence of the N_2O triple point in the regions measured. Since the boiling point, 184.67 K, and the triple point, 182.33 K, of a N_2Obulk phase at a pressure, 1 atm, are very close to each other, the related two phase transitions can be observed clearly from the isosteres measured as N_2O concentration exceeds the sorbent-saturation capacity (the access amount dosed becomes bulk liquid and/or solid phases). This particular feature is obvious from the r.h.s. isosteres in Figure 4. From the specific slopes of the two segments of the "isostere" for the highest concentration, *i.e.*, at 8.6133 mol/kg, the latent heat of evaporation, 16.55 kJ/mol at 184.67 K, and the latent heat of fusion, 6.54 kJ/mol at 182.33 K, were calculated. These quantities agree well with handbook data [62]. There is a transition region between these two straight segments of isosteres.

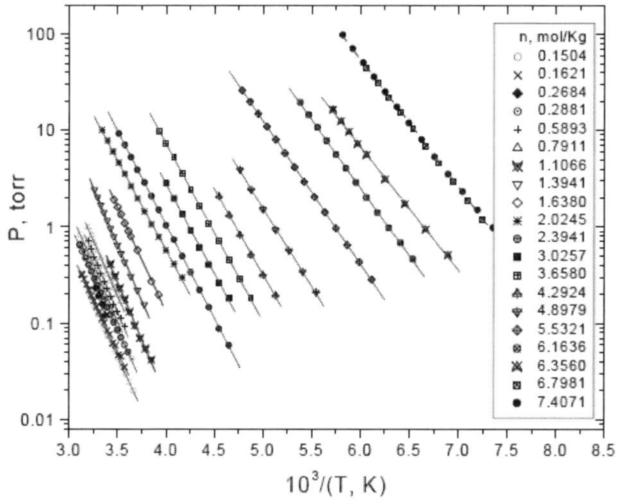

Figure 3. Sorption isosteres of CO_2 on NaLSX beads, FAU-I (the notation * refers to the "isostere" of CO_2 sublimation).

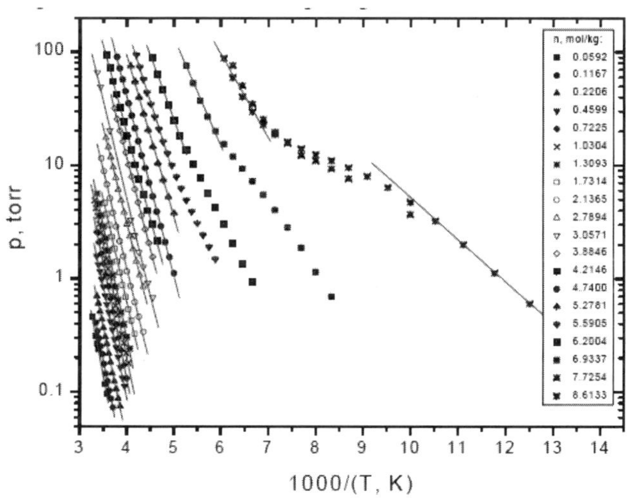

Figure 4. Sorption isosteres for N_2O on NaLSX beads at phase concentrations indicated.

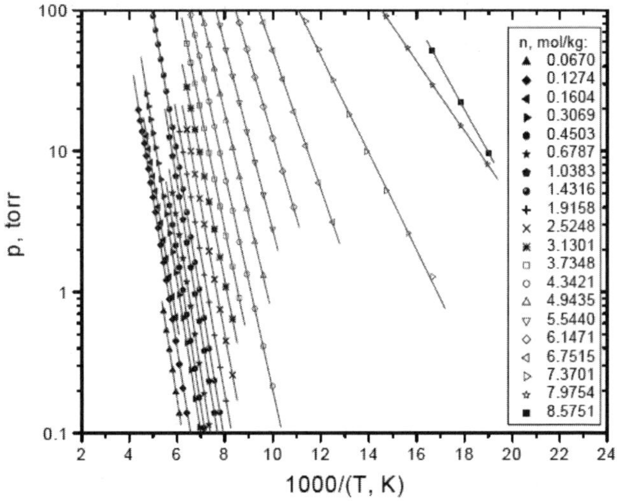

Figure 5. Sorption isosteres of N_2 on NaLSX beads at phase concentrations indicated.

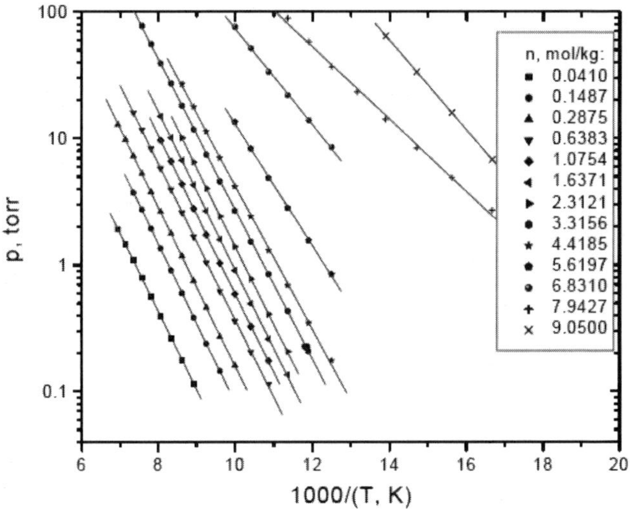

Figure 6. Sorption isosteres of O_2 on NaLSX beads, FAU-I, beads at phase concentrations indicated.

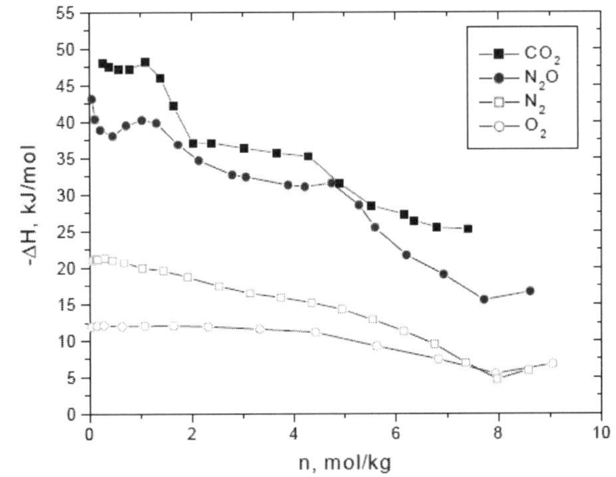

Figure 7. Concentration dependence of isosteric sorption heat of CO_2, N_2O, N_2 and O_2 on NaLSX zeolite beads.

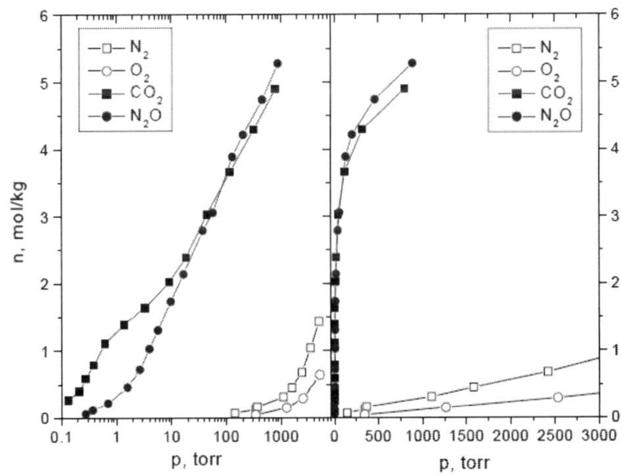

Figure 8. Sorption-isotherm sections calculated from SIM sorption-heat data for N_2, O_2, CO_2 and N_2O on NaLSX zeolite beads at 298 K in various pressure scales.

Dependences of values - ΔH (= *qst*) on sorption-phase concentrations, *n*, for CO_2, N_2O, N_2 and O_2, referred to the crystalline NaLSX phase, are shown in Figure 7. The value *qst* at very low values *n* for CO_2, N_2O, N_2, and O_2 on NaLSX zeolite amount to *c*. 48, 41, 21 and 12 kJ/mol, respectively. The difference between CO_2 and N_2O for values, *n*, between about 0 and 1 mol/kg is less than 7 kJ/mol, but this difference is diminished as *n* increases. Characteristic differences in *Gibbs* free sorption energy between CO_2 and N_2O determine the behavior of their sorption isotherms. It can be understood that since NaLSX exhibits sorption heats very much higher over the concentration ranges for both CO_2 and N_2O, compared to those for N_2 and O_2, the ability of this material is outstanding to remove CO_2 or N_2O from air, *i.e.*, N_2 and O_2, in related purification processes. This is exemplified by Figure 8, which shows sorption-isotherm sections for the various gases calculated from thermodynamic parameters obtained by SIM.

Favorable sorption properties of NaLSX towards CO_2 and N_2O are obvious, compared with those of that material with regard to N_2 and O_2. This makes NaLSX an outstanding sorbent for the pre-purification of air upfront cryogenic air separation units for the production of N_2 and O_2 [40]. Enhancement of favorable sorption properties towards CO_2 and N_2O can be achieved by cation exchange Na^+ *vs.* Ca^{2+} of the basic LSX phase. As a result of this, N_2O can be sorbed preferentially over CO_2 at sufficiently low phase concentrations [63] that exist under conditions of air pre-purification upfront its cryogenic distillation. This is demonstrated by Figure 9 that shows concentration dependences of standard *Gibbs* free sorption energies for CO_2 and N_2O on NaLSX and CaLSX zeolites, as they were derived from SIM data. That feature of cation-exchanged LSX sorbents could be shown to be useful for the removal of N_2O from air in the presence of CO_2 and light hydrocarbon gases as well, *cf.*, Table 1.

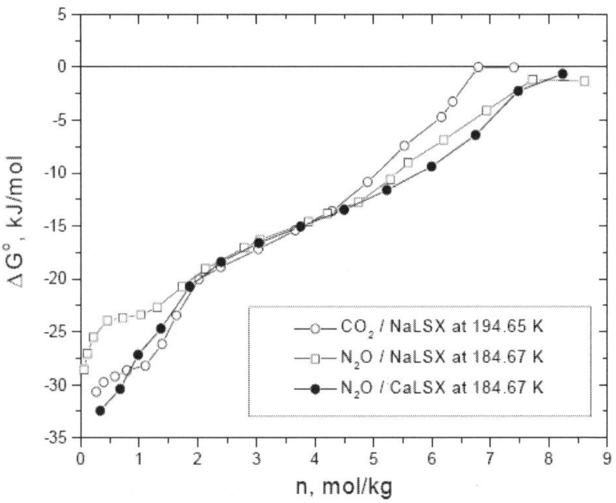

Figure 9. Standard *Gibbs* free sorption energies for CO_2 and N_2O on NaLSX and CaLSX zeolites.

Purification efficiency of such a sorbent with regard to N_2O is demonstrated by some data presented in Table 1.

Table 1. Trace-removal performance of BOC TSA PPU sorbent at 0.1 ppm CO_2 breakthrough.

Trace component	Formula	Feed / ppm	% Removal
Nitrous oxide	N_2O	0.35	100
Ethane	C_2H_6	1.4	85
Propane	C_3H_8	1.6	98
Ethylene	C_2H_4	1.5	100

4.2. *Sorption Heats of Carbon Dioxide on NaLSX, NaX and DAY Zeolites*

Sorption isosteres and sorption thermodynamic data for CO_2 on specific FAU zeolite modifications, NaLSX and NaX, *i.e.*, FAU-I and FAU-II (*cf.*, Table 2 in [30]), will be compared here with related data obtained for a DAY zeolite, *viz.*, dealuminated sub-type of the FAU-framework species. Figure 10 shows sorption isosteres measured for a DAY sample with a framework elemental Si/Al ratio of *c*. 56, supplied by *Degussa*, Germany.

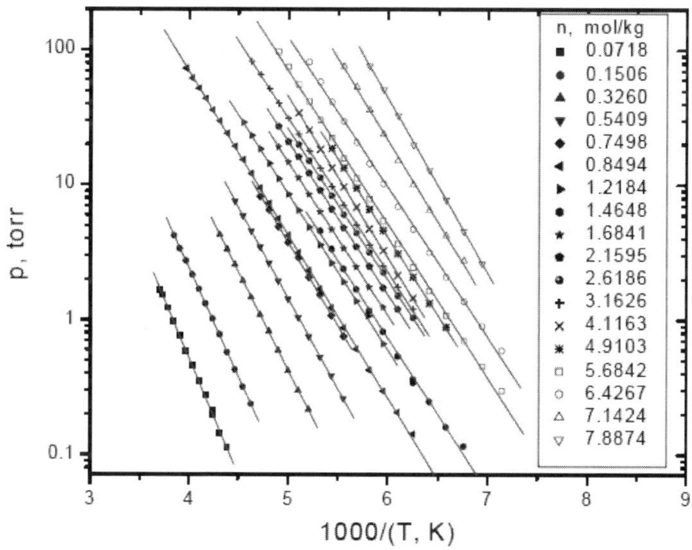

Figure 10. Sorption isosteres for CO_2 on *Degussa* DAY zeolite crystals at phase concentrations indicated.

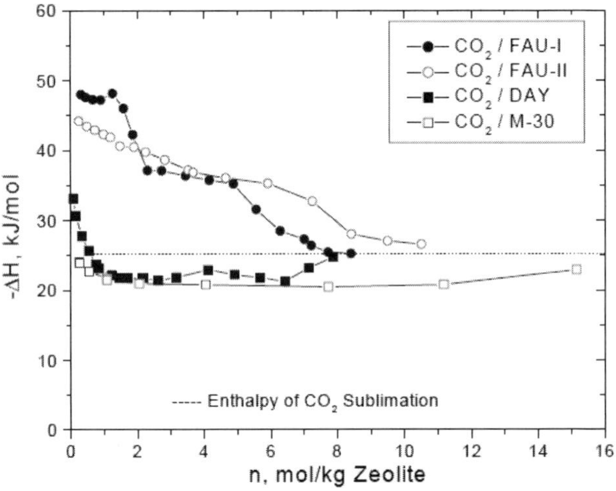

Figure 11. Concentration dependences of isosteric sorption heats for CO_2 on NaLSX (FAU-I), NaX (FAU-II), DAY zeolites and *Osaka Gas* carbonaceous sorbent M-30.

The isosteric sorption heats derived therefrom are shown in Figure 11 together with those for CO_2 sorption by both NaLSX and NaX, over the entire concentration ranges up to micropore saturation for the three systems. In addition to those data, isosteric sorption heats are shown for CO_2 on M-30, an *Osaka Gas* super-activated micro-mesoporous carbon material.

The concentration dependences of *qst* show several remarkable features: (*i*) upon saturation, the sorption heat for all materials reaches the value characteristic of CO_2 sublimation; this also indicates limiting values of sorption-phase saturation for the various materials; (*ii*) the isosteric heats on NaLSX and NaX proceed well above the heat of sublimation over the entire concentration range, and it approaches the latter at saturation only (peculiarities were discussed in [30]); (*iii*) the plateau for NaLSX at concentrations below *c.* 2 mol/kg could be referred, most probably, to sorption interaction between CO_2 molecules and Na+ cations of the FAU; (*iv*) sorption of CO_2 by DAY and M-30 follows a very similar energetic pattern: residual amounts of specific sorption sites that exhibit a somewhat higher sorption heat at very low values, *n*, and subsequent, almost identical curve courses, *qst vs. n*, below the sublimation heat of CO_2; (*v*) interaction between CO_2 and the intracrystalline "silica-like" surface of DAY as well as the intraporous carbon surface of M-30 seems to be close, which may be an interesting finding *per se* to be further dealt with; (*vi*) the saturation capacity for M-30 exceeds that of DAY by a factor of about 2.

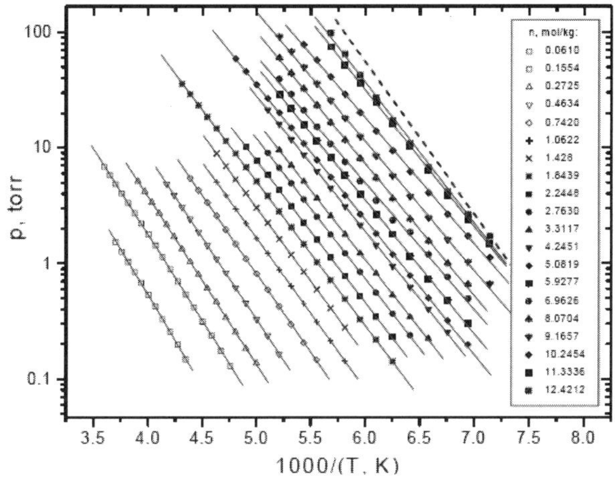

Figure 12. Sorption isosteres for CO_2 on *CarboTech* D 47/2 activated carbon.

4.3. *Sorption Heats of Carbon Dioxide on Carbonaceous Sorbents*

Sorption isosteres were investigated for CO_2 on a series of carbonaceous sorbents, specifically on materials D 47/2, D 55/2 and DGK that were kindly provided by *CarboTech*, Germany. These materials differ in their degree of activation (as manufacture step) and, thus, in their sorption capacity for CO_2, especially in the micro-mesoporous range. As an example, sorption isosteres for CO_2 on the D 47/2 sorbent are shown in Figure 12.

The sorption isosteres cover entire sorption-phase concentration ranges, from *c*. 0.06 mol/kg to *c*. 12.4 mol/kg. The sorption isosteres determined appear to be linear within the experimental conditions, indicating that no sorption-phase transition occurs in the system.

Sorption heats derived from the CO_2 isosteres shown in Figure 12, and those derived from similar plots for the other *CarboTech* materials as well as those for the M-30 sorbent are reproduced in Figure 13.

Although the four carbons show different sorption-saturation capacities for CO_2, similar concentration dependences of *qst* exist among these materials. Samples D 47/2 and DGK show nearly identical saturation capacities, whereas D 55/2 has a CO_2 saturation capacity less by *c*. 40 % compared to that for the two other materials of *CarboTech* origin. The specific behavior of the CO_2 /M-30 system has already been discussed elsewhere [30]. These differences in sorption thermodynamics lead to different sorptionequilibrium isotherms for CO_2, which cannot be shown here due to lack of space.

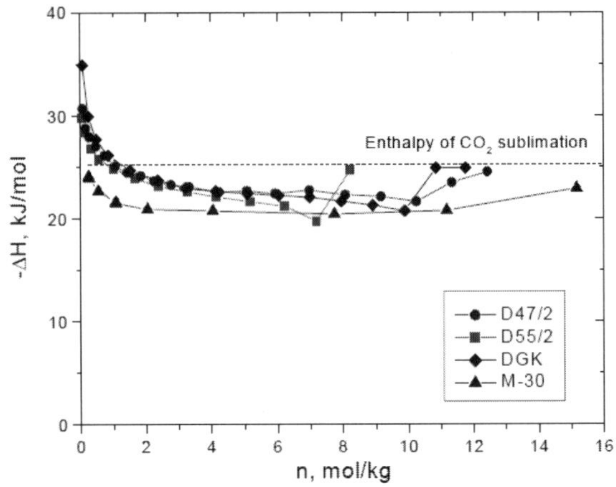

Figure 13. Concentration dependences of isosteric sorption heats of CO_2 on various carbonaceous sorbents: D 47/2, D 55/2, DGK from *CarboTech*; M-30 from *Osaka Gas*.

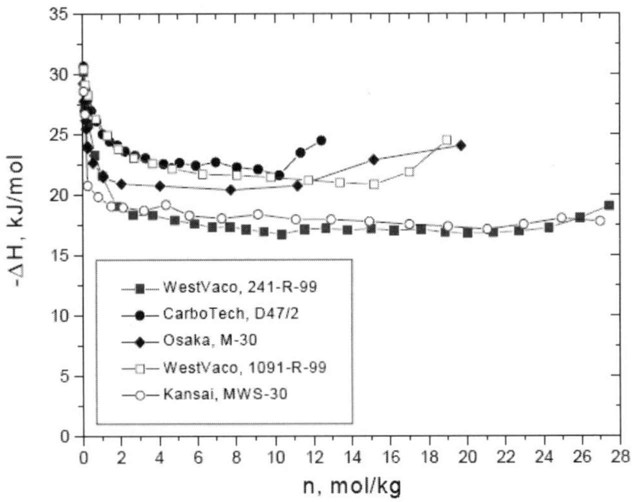

Figure 14. Concentration dependences of isosteric sorption heats for CO_2 on carbon sorbents: D 47/2 from *CarboTech*, Germany; M-30 from *Osaka Gas*, Japan; MWS-30 from *Kansai Coke & Chemicals (KCC)*, Japan; 1091-R-99 and 241-R-99 from *Westvaco*, USA.

Another comparison of concentration dependences of isosteric sorption heats of CO_2 on a series of up-to-date carbonaceous sorbents with highest sorption capacities - as they were determined by SIM -, is given in Figure 14 (more details will be given in [64]). Compared to the hard-coal based material D 47/2 that shows highest differential enthalpy changes for CO_2, the latter thermodynamic quantity decreases with increasing overall sorption capacity of the other materials. Proper sorbent tailoring with regard to differential sorption heat and sorption capacity for CO_2 may lead to an *optimum* integral ad(de)sorption heat, which would be relevant for adsorptive warming or desorptive cooling of fluids in closed containers, *e.g.*, of baby food and liquid beverages, respectively, the latter having been suggested, for example, in [65].

Related attempts had been made to calculate integral sorption heats of CO_2 on all materials investigated, *e.g.*, on D 47/2, by means of eq. (23) where the isosteric sorption heats (differential quantities) were used to calculate integral quantities over defined ranges of sorption-phase concentration, n, *e.g.*, between its limits $n = 0$ and $n = \mathrm{n}$,

$$q_{\mathrm{int}} = \frac{1}{n}\int_0^n q_{st}\,dn. \tag{23}$$

Utilizing eq. (23) for the specific purpose, one should have in mind that – in this case – the integral heat of ad/desorption used does not represent a single-phase property but that of an equilibrium between two phases in a sense that should be imagined as moving from one isotherm to another when moving from one concentration to another, $n_1 \Rightarrow n_2$, as a result of pressure changes in the system, *viz.*, $p_1 \Rightarrow p_2$, which is connected with a finite value of mechanical work executed. Thus, the mechanic work does play a role for an integral sorption heat, *cf.*, [66].

An integral de(ad)sorption heat as calculated for the CO_2/D 47/2 system, viz., 122 J/g, cf., Figure 15, would allow for a certain cooling (warming) of a liquid in close contact with the sorbent container, presupposing that the CO_2 equilibrium pressure over the sorption phase in the container changes from 20 to 1 bar, and temperature from 25 to 10 °C. Cooling efficiency could be nearly doubled by using other materials, e.g., M-30 [67], or those of KCC and Westvaco, that are, however, quite expensive. A comparison of desorptive cooling (warming) efficiency between various materials based on SIM data and directly measured high-p isotherms for CO_2 equilibria is shown in Figure 16. The influence of the pressure envelope on the efficiency is obvious.

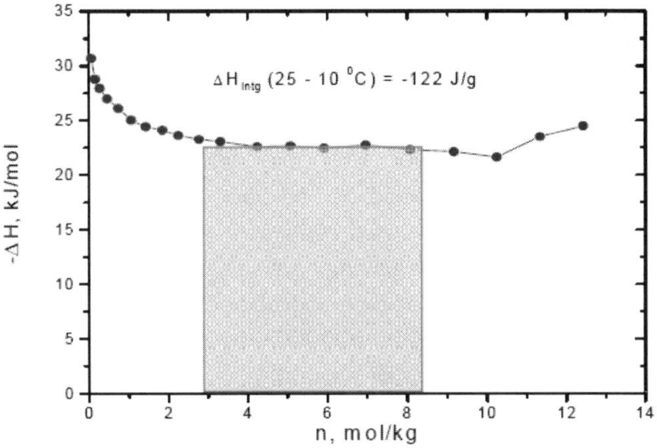

Figure 15. Determination of integral sorption heat exemplified for CO_2 on *CarboTech* D 47/2 material over the parameter ranges (*T*, *p*): (25 oC, 20 bar) to (10 oC, 1 bar).

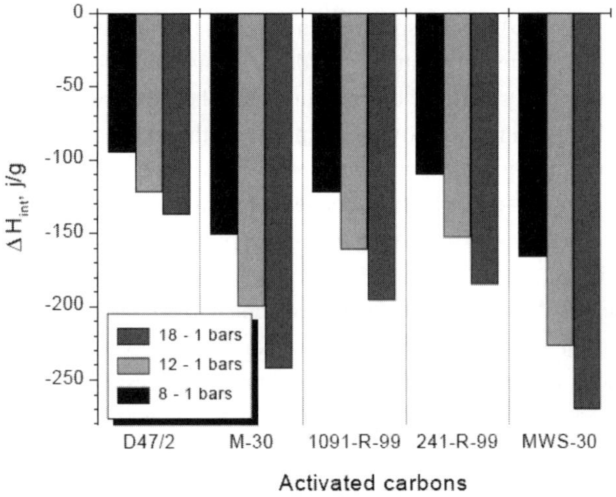

Figure 16. Comparison of pressure envelopes for de(ad)sorptive cooling (warming) based on integral sorption heats of CO_2 for various carbonaceous sorbents.

4.4. *Sorption Heats of Nitrogen on LiLSX and CaA Zeolites*

Sorption-isosteric heats determined by SIM over full concentration ranges can be analyzed to identify, quantify and distinguish between the strengths of sorption sites in nanoporous sorption systems. Figure 17 shows the concentration dependences of isostericsorption heats of N_2 and O_2 on zeolites CaA (Ca ion

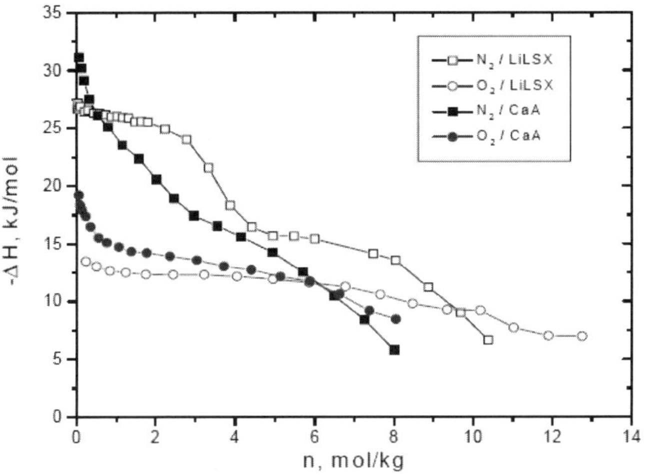

Figure 17. Isosteric sorption heats for N_2 and O_2 on LiLSX and CaA zeolites.

content≅ 97 %) and LiLSX (Li^+-ion content≅ 99 %), from which the following main conclusions can be drawn: (*i*) values of initial isosteric sorption heats for N_2 and O_2 on CaA zeolite are by *c.* 5 kJ/mol higher than those on LiLSX, which indicates that interactions of N_2 and O_2 molecules with Ca^{2+}-ionsites in CaA zeolite are stronger than those with Li^+-cation sites in LiLSX zeolite; (*ii*) the Li^+-ion sorption sites in LiLSX are energetically less heterogeneous than the Ca^{2+}-ion sorption sites in CaA for both N_2 and O_2 molecular sorption; (*iii*) compared with CaA, LiLSX zeolite provides energetically more strong and nearly homogeneous sorption centers for N_2 at loadings up to *c.* 2 mol/kg; (*iv*) LiLSX shows a weaker sorption potential for O_2 than CaA does; the difference in sorption heats between N_2 and O_2 on LiLSX is significantly larger than that on CaA, which results in much higher N_2 sorption selectivity over O_2 on LiLSX than on CaA; (*v*) the sorption-saturation capacities in LiLSX are larger than those in CaA, *i.e.*, the concentration dependences for N_2 and O_2 in LiLSX extend much far to the right; (*vi*) after approaching and finally exceeding the sorption-saturation capacities, the heats for bulk liquid-gas phase transitions were measured, *i.e.*, 6.82 kJ/mol for O_2 and 5.58 kJ/mol for N_2.

Figure 18 shows sorption isotherms of N_2 and O_2 on LiLSX zeolite at 25℃, which were obtained by molecular simulations and microbalance experiments, along with isotherms calculated from sorption-thermodynamic functions obtained by SIM. Obviously, these isotherms are in good agreement with each other.

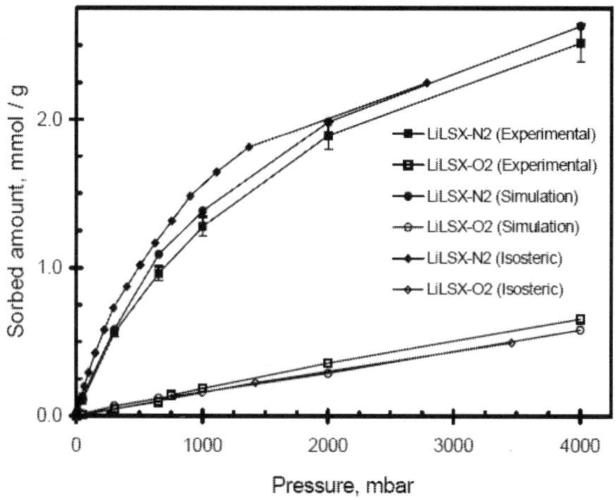

Figure 18. Sorption isotherms for N2 and O2 on LiLSX at 25 oC from SIM, microbalance experiments and *Monte Carlo* simulations.

Differential sorption heats were provided by *Monte Carlo* simulations of sorption processes, *viz.*, from the slope of curves obtained by plotting values of total potential energy against sorption-phase concentration. The isosteric sorption heat can then be calculated by adding the mechanical-work term to the differential sorption heat assuming that the gas is ideal and the sorption phase is denser than the gas phase.

Simulated isosteric sorption heats for N_2 and O_2 for a LiLSX structure that contains Li^+ ions with a modified charge, + 0.95, are plotted against the sorption-phase concentration in Figure 19 along with the experimental data for comparison, *cf.*, ref. 60.

As expected, the isosteric sorption heat decreases gradually with increasing sorptionphase concentration, for both simulated and experimental data. However, the simulated values of isosteric sorption heat are somewhat higher than the experimental data. This difference increases with sorption-phase concentration and amounts to *c.* 2 kJ/mol, at the most. Interestingly, an almost analogous qualitative and quantitative picture resulted from comparative isosteric and calorimetric studies of concentration dependences of isosteric sorption heats for N_2 and O_2 on identical CaA samples, which was performed independently [36,37]. In the LiLSX case, however, simulated values of the isosteric sorption heat for O_2 are slightly lower than the experimental data. Since the *Coulomb*-type interactions between O_2 molecules and cations are very weak, the sorption heat of O_2 is much lower than that of N_2, *cf.*, below.

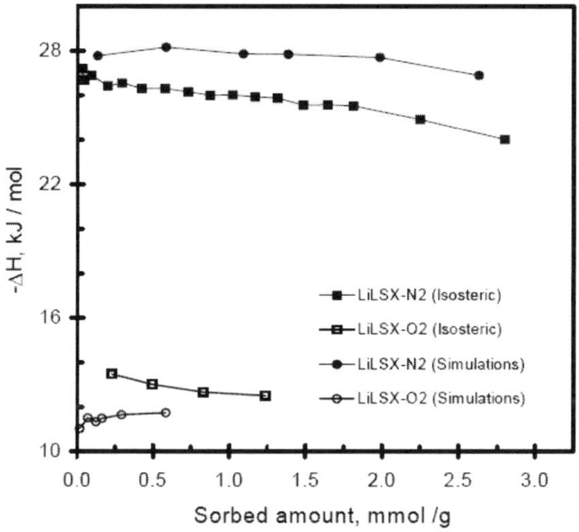

Figure 19. Isosteric sorption heats for N_2 and O_2 on LiLSX from SIM experiments and *Monte Carlo* simulations.

4.5. *Sorption Heats of Nitrogen and Oxygen on Li,RE-LSX Zeolite for Oxygen PVSA*

Zeolite Li,RE-LSX for O_2 PVSA used herein was a representative sample of large-scalemanufacture batches, *i.e.*, beads. It was prepared in accordance with [39,68,69]. The Si/Al ratio of its FAU framework was≅ 1.01. Concentrations of ions of Li^+ and of the trivalent metals in the Li,RE-LSX material corresponded to those of BOC-proprietary compositions [39] with Li^+ ion concentrations being outside the range claimed in [70]. Residual sodiumplus-potassium ion levels of all Li,RE-LSX specimens were less than *c*. 2 %, on an equivalent's basis. Sorption results for beaded samples were corrected for binder content. Homogeneous distributions of Al, Si, Li, Na and trivalent metals, *etc*., were proven by Time-of-Flight SIMS studies performed on randomly chosen Li,RELSX-bead samples by means of a Physical Electronics instrument, Phi-Evans TFS-2000, with a $^{69}Ga^+$ liquid metal-ion gun as primary ion beam, over analysis regions, (200 μm)^2, and, (240 μm)^2, of "microtome-like" prepared bead surfaces [71], *cf*., Figure 20. There were the following main results:
(*i*) distribution of elements over cross-sectional analysis areas is homogeneous, within the accuracy and resolution limits of the TOF-SIMS technique; (*ii*) in accordance with proprietary methods [39,68,69] of preparation of the materials, no gradients in concentration of ions, particularly Li^+ and La^{3+} exist, which holds

for bulk and edge areas of any zeolite beads looked at; (*iii*) a certain amount of La exists as LaO species that could be located tentatively in the FAU supercages over entire bead regions; (*iv*) no evidence of beadcomposition-alien surface "skins", patches of deposited layers or non-zeolitic phases were found in any TOF-SIMS experiments. The gradient-free distribution of ions in Li,RELSX-zeolite composites is important to ensure high PVSA performance of O_2 production.

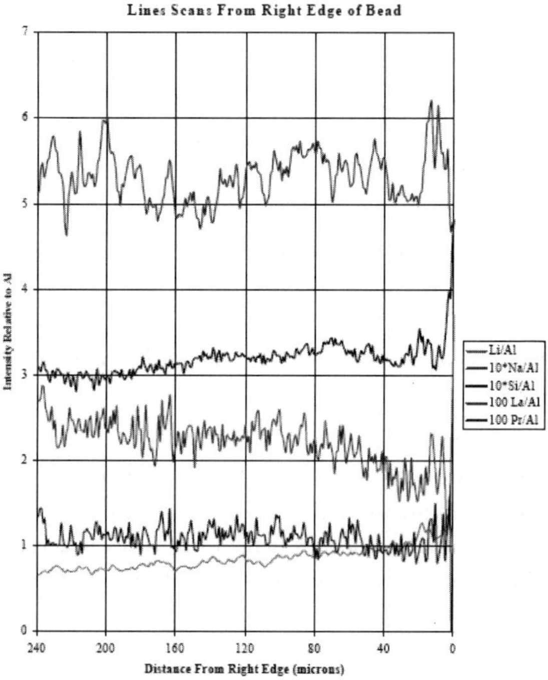

Figure 20. SIMS line scans for distributions of elements in Li,RE-LSX zeolite beads; sample 2016.

Isosteric sorption heats of N_2 and O_2 obtained by SIM for Li,RE-LSX zeolite are shown in Figure 21 as dependences on sorption-phase concentration, *n*. The plots - ΔH *vs. n* for N_2 and O_2 exhibit three characteristic ranges: (*i*) at *c.* $n \lesssim 3$ mol/kg, which reflects specific interactions of N_2 and O_2 quadrupoles with Li+ ions that may, in principle, occupy energetically different extra-framework sites; (*ii*) at *c.* ($3 \lesssim n \lesssim 6$) mol/kg, which is governed by mostly non-specific *van der Waals*-type interactions, between gas molecules and the zeolite framework, and intermolecular interactions; (*iii*) at *c.* ($6 \lesssim n \lesssim 9$) mol/kg, *i.e.*, for n approaching and exceeding the saturation capacity, *n* max , that represents micropore filling processes. Although the two systems exhibit three similar regions, the strength of those interactions is significantly higher for N_2 than for O_2. The specific interaction between N_2 and Li^+ ions is about three times stronger than that between O_2 and Li^+ ions, since this type of specific interactions is proportional to the values of quadrupole moments, *c.* 0.3 $Å^3$ for N_2 and 0.1 $Å^3$ for O_2. This feature implies that any zeolite modification to increase its ability for specific interactions would improve N_2 sorption over that of O_2 by about a factor 3, thus, increasing strongly the separation selectivity of N_2 over O_2. At $n \to n_{max}$, the heat effects approach those for liquefaction (evaporation) of the gases, i.e., 5.58 kJ/mol for N_2 and 6.82 kJ/mol for O_2. The sorption-saturation capacities n_{max} amount to c. (8 ÷ 9) mol/kg for both gases on the sorbent given. A comparison of sorption heats of N_2 between Li,RE-LSX and LiLSX zeolites is presented in Figure 22. Although the patterns are similar, they differ significantly at $n \lesssim 3$ mol/kg, *viz.*, the specific N_2-ion interaction for Li,RE-LSX exceeds that for LiLSX by *c.* (4 ÷ 5) kJ/mol, and then again at $n \approx (4 \div 8)$ mol/kg.

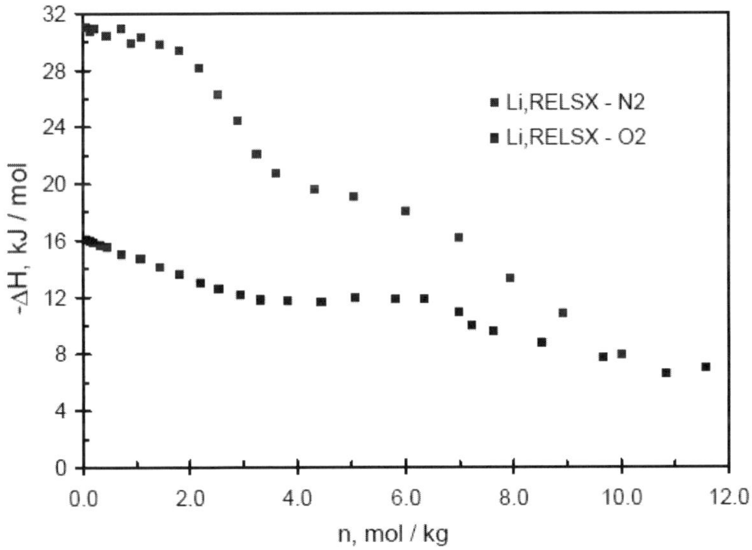

Figure 21. Concentration dependences of isosteric sorption heats for N_2 and O_2 on Li,RE-LSX zeolite.

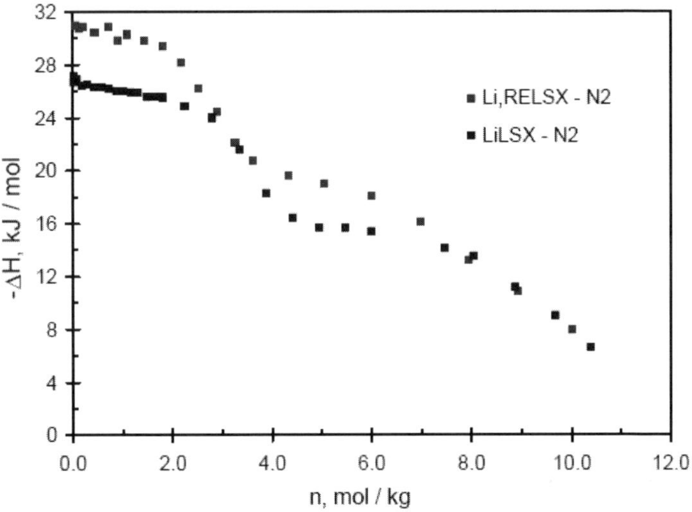

Figure 22. Comparison of concentration dependences of isosteric sorption heats for N_2 on Li,RE-LSX and LiLSX zeolites.

The former difference could be addressed by *Monte Carlo* simulation of N_2 interaction with Li,La-LSX and LiLSX systems, *cf.*, [71]. On the other hand, although the initial sorption heat of O_2 is higher for Li,RE-LSX compared to LiLSX, its difference for the two materials is smaller than that for N_2 sorption.

The Cation-locator module of the Accelerys Cerius2 software package was used to position the Li^+ and trivalent metal ions based on known XRD structure data for LiLSX. Simulated sorption isotherms are in excellent agreement with experimental data. Simulations also predict that Li^+ and La^{3+} ions in sodalite cages and Li^+ ions at sites SII in FAU supercages do not participate in the sorption process. La^{3+} ions at sites SII attract N_2 molecules compensating the loss of a number of accessible Li^+ ions. The presence of La at SII site facilitates bridging La^{3+} ion at SII and Li^+ ion at SIII/SIII' sites by N_2 molecule, *cf.*, Figure 23. This phenome-non leads to additional distinct sorption sites with stronger interaction energy, which correlates to the finding of higher heats of N_2 sorption obtained by SIM experiments, and a more heterogeneous surface in Li,RE-LSX compared to that in LiLSX zeolite.

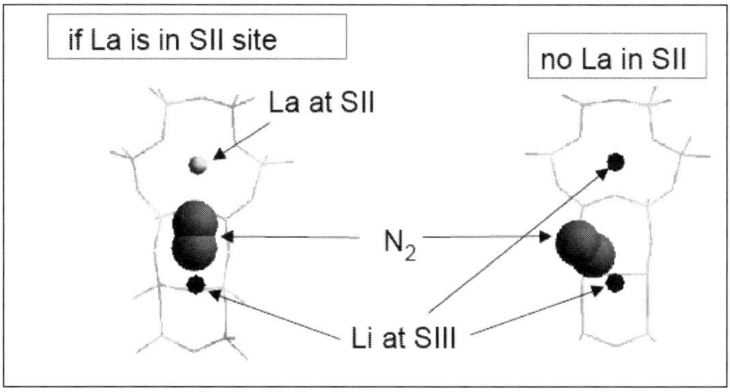

Figure 23. Geometry of sorbed N_2 molecule in Li(84)La(4-SII)-LSX and Li(96)LSX at the end of *Monte Carlo* sorption simulation; 298 K.

4.6. *Sorption Heats of Nitrogen - Oxygen Mixtures on Li,RE-LSX Zeolite*

SIM experiments for binary N_2-O_2 mixtures on Li,RE-LSX zeolite were performed in conjunction with single-component investigations. Mixture measurements are exemplified by isosteres for a sorption-phase composition of 80 % N_2 and 20 % O_2 as shown in Figure 24, over the entire concentration ranges for zeolitic intracrystalline void volume up to filling secondary pore volumes of the beads, as also observed from isosteres. In those representations, each line of symbols is one isostere measured at the respective sorption-phase concentration. As the latter concentration approaches saturation, the orresponding isostere approaches the sublimation curve of either the single component or the binary mixture. The coincidence between isostere and sublimation curve beyond saturation capacity proves that isosteric measurements were correct and thermodynamically consistent.

Differential sorption enthalpy as function of sorption-phase concentration, *cf.*, Figure 25, shows different profiles for pure N_2, pure O_2 and their mixtures on Li,RE-LSX. The stepwise and well-defined sorption energies of the single-component systems as dependencies on concentration, are discussed above. For the N_2 - O_2 binary mixture at sorption-phase composition 80 % N_2 and 20 % O_2, however, it is surprising that the isosteric sorption heat for the binary mixture is very much close to that of pure O_2.

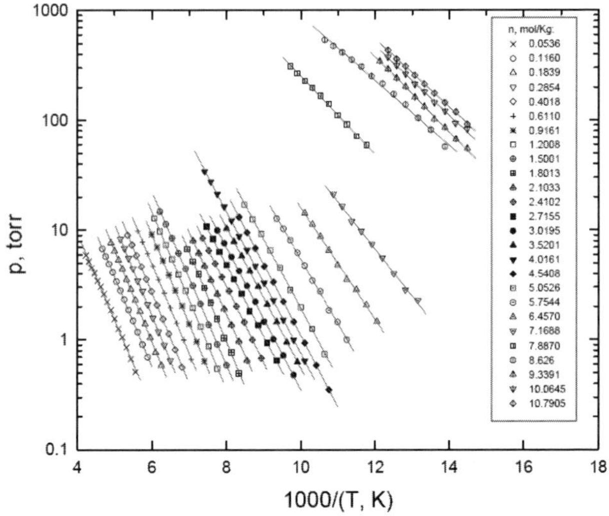

Figure 24. Sorption isosteres of binary N_2 - O_2 mixtures on Li,RE-LSX zeolite at sorption-phase composition of 80 % N_2 and 20 % O_2.

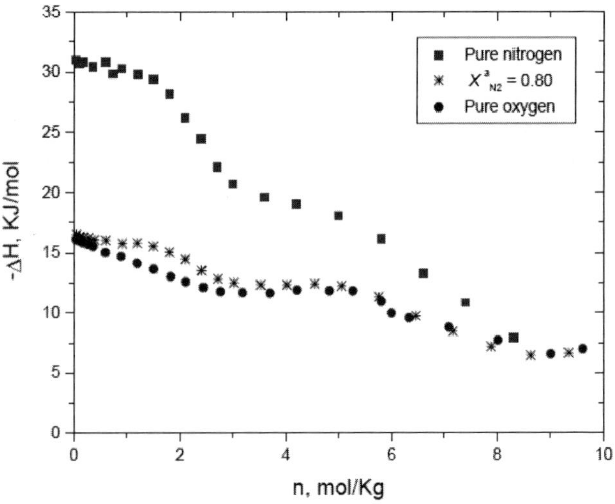

Figure 25. Concentration dependences of isosteric sorption heat for N_2, O_2 and their binary mixtures at sorption-phase composition, 80 % N_2 and 20% O_2, on Li,RE-LSX zeolite.

The standard sorption entropy, $\Delta S°$, for N_2, O_2 and their binary mixtures, which is referred to the standard-state gas pressure, 760 torr, and calculated as function of sorption-phase concentration, *cf.*, Figure 26, also shows significantly different profiles. A remarkable entropy loss for sorbed molecules compared to the standard gas phase, occurs over the entire concentration range. The change, $\Delta S°$, varies between *c.* -30 and -120 J/mol K. From an entropic point of view, N_2 molecules are more strongly confined in zeolitic micropores, compared with O_2 and N_2 - O_2 mixtures. For the well-defined heterogeneous sorbent, wave-like sorption-entropy dependences for N_2, O_2 and their binary mixtures on concentration are found. This pattern corresponds to that of the differential sorption enthalpy as described above, *i.e.*, it is characteristic of a model for occupying several groups of energetically equivalent sorption sites in the sequence of their interaction energies. A wave-like profile in entropy change is in excellent agreement with computer-simulation results for a heterogeneous surface [72].

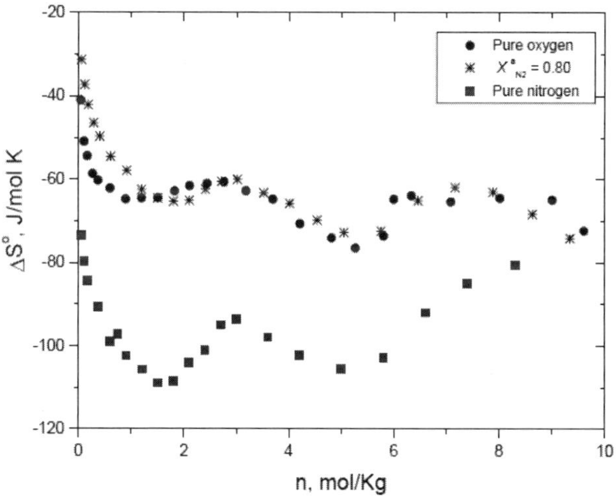

Figure 26. Concentration dependences of standard sorption entropy for N_2, O_2 and binary mixtures at sorption-phase composition, 80 % N_2 and 20% O_2, on Li,RE-LSX zeolite.

Gibbs free energy characterizes the natural tendency of a system to its spontaneous change. Dependences of standard *Gibbs* free sorption energies, $\Delta G°$, on sorption-phase concentration in Li,RE-LSX as referred to the boiling temperatures and 760 torr are shown in Figure 27.

In the three systems, $\Delta G°$ changes from negative values to zero as sorption-phase concentration increases and exceeds saturation capacities. This demonstrates thermodynamic consistency of experimental data. The larger negative values of $\Delta G°$ in cases of N_2 sorption on Li,RE-LSX indicate a stronger exothermic sorption process compared to those of O_2 and mixtures, whose $\Delta G°$ data amounts to only about half of that for N_2 at initial concentration.

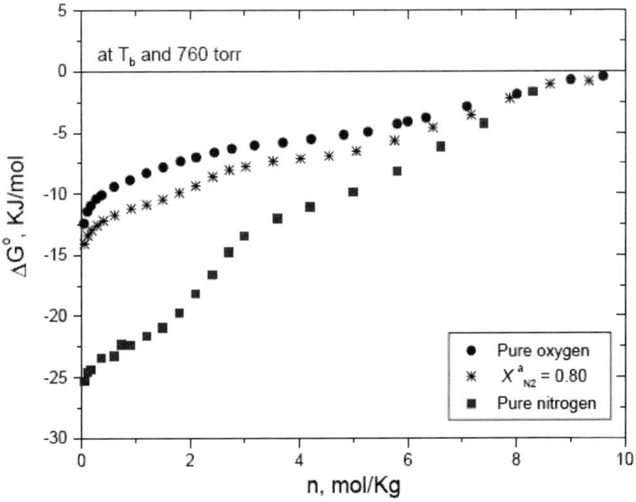

Figure 27. Concentration dependences of standard *Gibbs* free sorption energy for N_2, O_2 and binary mixtures at sorption-phase composition, 80 % N_2 and 20 % O_2, on Li,RE-LSX zeolite, referred to the boiling temperatures and 760 torr.

As described above, with specific reference to the AST approach, experimental isosteric data, specifically, standard *Gibbs* free sorption energy as concentration dependences allow for both interpolation and extrapolation of sorption isotherms for any physically meaningful regions of temperature and pressure. In a first step, the initial values of sorption enthalpy and entropy for single components and binary mixtures were obtained by fitting the thermodynamic functions with the polynomial equations (11-12). The initial *Gibbs* free sorption energy changes were then calculated *via* the concentration dependences of sorption enthalpy and entropy. The initial thermodynamic values for the *Henry* region as function of sorption-phase composition on Li,RE-LSX are shown in Figures 28-30.

The initial isosteric sorption heats for all mixture compositions up to that of 90 % N_2 are surprisingly close to that of pure O_2, and there is a sharp increase in the initial heat as sorption-phase composition approaches that of pure N_2.

The initial standard entropy change obtained corresponds to the enthalpy change that shows a slight increase as sorption-phase composition increases to *c.* 90 % of N_2, and then sharply decreases to the value for pure N_2. As sorption-phase concentration reduces to zero, *i.e.*, towards the *Henry* region, the sorption phase should behave like an ideal solution. The initial *Gibbs* free sorption energy data as function of sorption-phase composition at 298 K are compared with those from IAST prediction from single-componentdata in Figure 30. A reasonable agreement is achieved between these two data sets considering certain errors in initial entropy values. Although there are sudden changes in composition dependences of initial enthalpy and entropy values, the initial *Gibbs* free sorption energy changes gradually from the value for pure O_2 to that of pure N_2, as it had been expected.

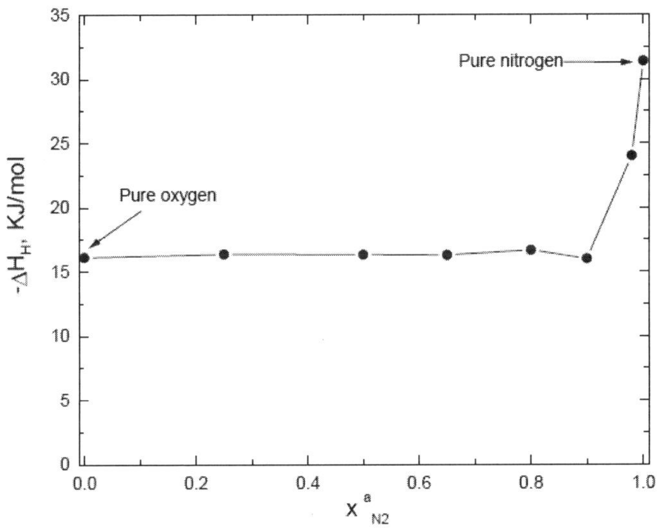

Figure 28. Initial isosteric sorption heat *vs.* sorption-phase composition for N_2 - O_2 mixtures on Li,RE-LSX.

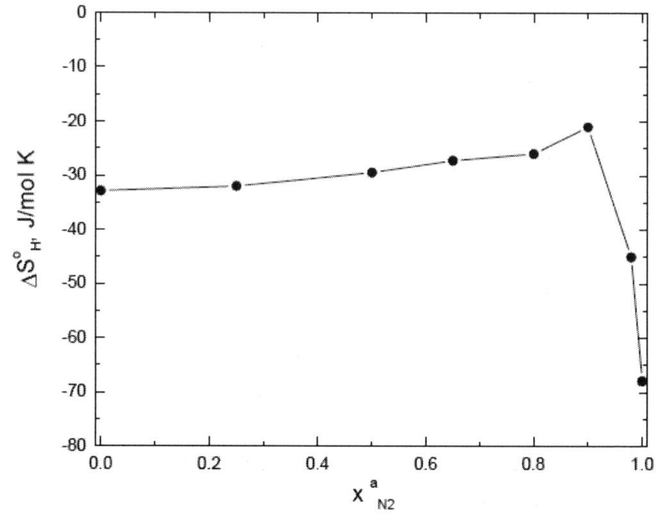

Figure 29. Initial standard sorption entropy *vs.* sorption-phase composition for N_2 - O_2 mixtures on Li,RE-LSX.

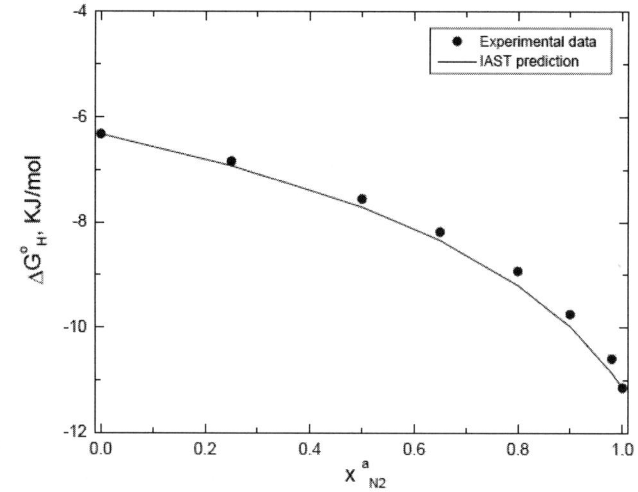

Figure 30. Initial *Gibbs* free sorption energy change *vs.* sorption-phase composition for N_2 - O_2 mixtures on Li,RE-LSX at 298 K.

The single-component thermodynamic data renders possible a prediction of mixture thermodynamic functions using solution thermodynamics and, thus, a precalculation of mixture sorption isotherms. An extended version of the method enables one to obtain directly partial values of thermodynamic quantities.

5. Conclusions

A modern version of the sorption-isosteric method has been shown to be a very useful tool for sorption-thermodynamic studies. Concentration dependences of thermodynamic functions over entire sorption-phase concentration ranges can be determined. During an isosteric measurement, fluid-component transfer between co-existing phases is kept to aminimum to ensure that isosteric conditions are maintained, and to accelerate equilibration between phases. Isostere "linearity" is assumed to occur, and its validity is discussed.

Measurements of full sets of sorption-thermodynamic data can be achieved reliably and rapidly with computerized control systems for high data accuracy. Correction for de(ad)sorption due to inherent temperature changes during SIM experiments can be made. Sorption-saturation values of a system can be assessed if its isosteres coincide withcharacteristic bulk-phase transition curves, *e.g.*, evaporation or sublimation curves. Phase transitions of the sorption phase can be observed directly from characteristic bending of isosteres. Sorption isotherms at any temperature and pressure that are physically meaningful, can be calculated from either concentration dependences of thermodynamic functions or directly from sets of isosteres.

SIM has been extended successfully to the investigation of sorption thermodynamics of multi-component mixtures. For the first time, it has allowed for determination of differential sorption heat and entropy data of ternary gas mixtures sorbed [21]. The method provides high-accuracy caloric data and allows for further development of fundamental knowledge of both experimental behaviors and related theoretical treatment. Some limitations to general utilization of SIM exist. So far, SIM is limited to nanoporous, *i.e.*, microporous and complex micro-mesoporous sorbents that are assumed - as a rule but not necessarily - to be inert during sorption processes. A small dead volume of the sorption system is a stringent prerequisite for utilization of the inherent high accuracy of SIM for equilibrium measurements. There are certain constraints in either low- or high-pressure regions, *viz*, equilibration, desorption rate, pressure-measurement accuracy, leak rate, and thermal-transpiration effects. The desorbed amount can be corrected for, for single components, but cannot be corrected for, for mixtures, due to practical reasons. SIM demands for a

T-gradient-free sorption cell, and it needs efficient gas circulation therein, especially for mixtures - demands, which were satisfied by sophisticated experimental arrangements. Corrections may be needed for deformation of microporoussorbents at high sorption-phase concentration to interpret results correctly.

A series of SIM data is compared with those from simulation experiments using *Monte Carlo* methods, and excellent agreement has been achieved. The energetic heterogeneity of sorbents due to specific interactions between molecules of various gases, *e.g.*, carbon dioxide, and specific sorption centers in zeolites, is quantified by characteristic concentration dependences of the thermodynamic functions.

SIM has been recognized nowadays as one of the important methods that lead to high-accuracy sorption-thermodynamic data, beside those of sorption calorimetry of various types and differentiation of sorption isotherms at constant sorption-phase concentration. Beyond any doubt, the method will contribute not only to further development of sorption separation and purification methods of direct industrial relevance as addressed in this paper, but also to elaboration of methods for pre-calculation of sorption equilibria of fluid mixtures based on single-component data as investigated, for example, by *Myers* and *Siperstein* [73], for further recognition of fundamental behavior of fluid-solid interface phenomena as developed by *Fomkin* [5], for finding structure-property relationships in heterogeneous catalysis as shown by *Mishin* [74], and for many other applications to come.

Acknowledgements

The author thanks Drs. Dongmin Shen, NJ, and Sudhakar R. Jale, CA, for their significant contributions to the work presented and great friendship during a decade of technical collaboration. He also acknowledges kindness and permanent support by Drs. Frank R. Fitch and Adeola F. Ojo, his former colleagues at *BOC PGS Technology*, Murray Hill, NJ.

References

1. (a) Avgul N.N., Kiselev A.V. and Poshkus D.P., *Adsorption of Gases and Vapors on Homogeneous Surfaces* (Russ.) (Chimija, Moscow, 1975) Chapter III. Kiselev A.V., Poshkus D.P. and Jashin, Ja.I., *Molecular Fundamentals of Adsorption Chromatography* (Russ.) (Chimija, Moscow, 1975) Chapters 3-6.

Kiselev A.V., *Intermolecular Interactions in Adsorption and Chromatography*(Russ.) (Vys'shaja Shkola, Moscow, 1986) pp. 149-160.
(b) Barrer R.M., *Zeolites and Clay Minerals as Sorbents and Molecular Sieves* (New York, Academic Press, 1978).
(c) Isirikjan A.A., in *Contemporary Problems of Adsorption Theory* (Russ.), ed. By V.V. Serpinsky, N.S. Poljakov and K.O. Murdmaa (Institute of Physical Chemistry, Russian Academy of Sciences, Moscow, 1995) pp. 72-84.

2. Rouquerol F., Rouquerol J., and Sing K.S.W., *Adsorption by Powders and Porous Solids, Methodology and Applications* (Academic Press, London, 1999).
3. Keller J.U., Rave H., Seelbach M. and Staudt R., in *Adsorption Science and Technology*, Proc. 2d Pacific Basin Conf. Adsorption Science and Technology, ed. by D.D. Do (World Scientific Publishing Co., Singapore, London) pp. 329-335. Keller J.U. and Staudt R., *Gas Adsorption Equilibria,* (Springer, Heidelberg, 2005).
4. Bülow M., Shen D., and Jale S., *Appl. Surf. Sci.* **196** (2002) 157-172.
5. Fomkin A.A., *Adsorption* **11** (2005) 425-436.
6. Serpinsky V.V. , *Colloquium*, Institut für Physikalische Chemie, Deutsche Akademie der Wissenschaften zu Berlin, Berlin-Adlershof, 1967.
7. Bering B.P., Žukovskaja E.G., Rachmukov B.Ch., and Serpinsky V.V., *Z. Chem. (Leipzig)* **9** (1969) 13-22.
8. Fomkin A.A. , Regent N.I. , Gusev V.Ju., Tkachenko S.G. , Eroshenko V.A. , and Serpinsky V.V. , *Izv. Akad. Nauk USSR, Ser. Chim.* 1989, 1386-1389.
9. Gusev V.Ju. and Fomkin A.A., *J. Coll. Interface Sci.* **162** (1994) 279-288.
10. Fomkin A.A. , in *Modern Theoretical Models of Adsorption in Porous Media* (Russ.), ed. by N.S. Poljakov and A.A. Fomkin (Inst. Phys. Chem., Moscow, 1999) p. 10.
11. Fomkin A.A., Pulin A.L, and Gusev V.Ju., in *Modern Problems of Theory of Adsorption and Synthesis of Sorbents* (Russ.), ed. by N.S. Poljakov and A.A. Fomkin (Inst. Phys. Chem., Moscow-Kljasma, 2000) p. 16.
12. Meinert G., Großmann A., Bülow M., and Schirmer W., in *Adsorbents, Their Synthesis, Properties and Utilization* (Russ.), ed. by M.M.Dubinin and T.G. Plačenov (Academy of Sciences of the USSR, Leningrad, 1971) pp. 164-169.
13. (a) Meinert G., *Das Adsorptionsgleichgewicht unter isosteren Bedingungen*, Dissertation: Dr. rer. nat. (Humboldt-Universität Berlin, 1969).
(b) Großmann A., *Über die Bedeutung halbempirischer Adsorptionstheorien bei der Gewinnung von Parametern der Einzelgasadsorption sowie ihre Anwendung auf die adsorptive*

Gemischtrennung, Dissertation: Dr. sc. nat. (Deutsche Akademie der Wissenschaften, Berlin, 1972).
(c) Thamm H. , *Experimentelle und theoretische Untersuchungen zur Charakterisierung der Adsorptionswechselwirkungen der C6-Kohlenwasserstoffe Benzol, Cyclohexan und n-Hexan am NaX-Zeolith*, Dissertation: Dr. rer.nat. (Zentralinstitut für PhysikalischeChemie, Forschungsbereich Chemie, Akademie der Wissenschaften der DDR, Berlin, 1975).
(d) Ortlieb H.-J., *Theoretische Behandlung der Adsorption von Normalparaffinen an zeolithischen Adsorbentien auf der Grundlage der statistischen Thermodynamik und Anwendung der Ergebnisse zur umfassenden Auswertung experimentellen Materials*, Dissertation: Dr. rer.nat. (Zentralinstitut für Physikalische Chemie, Forschungsbereich Chemie, Akademie der Wissenschaften der DDR, Berlin, 1986).

14. Stach H. and Fiedler K., *Sitzungsber. Akademie der Wissenschaften der DDR* Jahrg. 1981, No. 1/N, 43-66.
15. Stach H. and Fiedler K., *Sitzungsber. Akademie der Wissenschaften der DDR* Jahrg. 1981, No. 3/N, 86-104.
16. K. Fiedler, U. Lohse, J. Sauer, H. Stach, H. Thamm, and W. Schirmer, in *Proc. 5th Internat. Conf. Zeolites*, ed. by L.V.C. Rees (Heiden, London, 1980) pp. 490-500.
17. Blank H. , Bülow M. , and W. Schirmer, *Z. phys. Chem. (Leipzig)* **260** (1979) 395-400.
18. Lutz W., Bülow M. , Großmann A. , and W. Schirmer, *Chem. Techn. (Leipzig)* **31** (1979) 527-529; *Chem. Techn. (Leipzig)* **33** (1981) 136-138.
19. (a) Bülow M. , Wappler H.-J. , Piotrowska J., and Jaroniec M. , *J. Coll. Interf. Sci.* **85** (1982) 457-462.
(b) Bülow M., Jaroniec M., and Piotrowska J., *Thin Solid Films* **88** (1982) 373-379.
(c) Jaroniec M., Piotrowska J., and Bülow M., *Thin Solid Films* **106** (1983) 219-224.
20. Bülow M. and Lorenz P., in *Fundamentals of Adsorption* ed. by A.I. Liapis (Engng. Found., New York, 1987) pp. 119-128.
21. Bülow M., *Stud. Surf. Sci. Catalysis* **83** (1994) 209-215.
22. Bülow M. and Rees L.V.C. , *Adsorption and Diffusion on Zeolites*, Joint Research Project executed by The Central Institute of Physical Chemistry, Berlin, and The Imperial College of Science, Technology and Medicine, London, and sponsored by *The Academy of Sciences of the German Democratic Republic*, Berlin, and *The Royal Society of Great Britain*, London, under their *Agreement of Scientific Collaboration*; February 17, 1986 - October 3, 1990.

23. Graham P. , Hughes A.D. , and Rees L.V.C., *Gas Sep. Purif.* **3** (1989) 56-64.
24. Rees L.V.C., Brückner P., and Hampson J.A. , *Gas Sep. Purif.* **5** (1991) 67-76.
25. Hampson J.A. and Rees L.V.C. , *J. Chem. Soc. Faraday Trans.* **89** (1993) 3169-3176.
26. Yang Y. and Rees L.V.C. , *Micropor. Mater.* **12** (1997) 117 - 122.
27. Sircar S., *Ind. Eng. Chem. Research* **31** (1992) 1813-20.
28. Karavias F. and Myers A.L., *Langmuir* **7** (1991) 3118-3126.
29. Bülow M. and Shen D., *Direct Measurement of Sorption Isosteres for Gases on Microporous Solids*, in *Proc. Topical Conf. Sep. Sci. Technol.*, Part II, AIChE 1997 Annual Meeting, ed. by W.S.W. Ho and R.G Luo (Los Angeles, November 16-21, 1997) pp. 1150-1155.
30. Shen D. and Bülow M. , *Micropor. Mesopor. Mater.* **22** (1998) 237-249.
31. Bülow M. and Shen D., in *Fundamentals of Adsorption-6*, Proc 6th Internat. Conf. Fundamentals of Adsorption, ed. by F. Meunier (Elsevier, Paris, 1998) pp. 87-92.
32. Bülow M. and Shen D., in *Adsorption by Porous Solids*, Fortschritt-Berichte VDI, Reihe 3: Verfahrenstechnik; Nr. **555**, ed. by R. Staudt (VDI Verlag, Düsseldorf, 1998) pp. 120-131.
33. Shen D.and Bülow M., in *Proc. 12th Internat. Zeolite Conf.*, Baltimore, Maryland, 1998, ed. by M.M.J. Treacy, B.K. Marcus, M.E. Bisher, and J.B. Higgins (Materials Research Science, Warrendale, 1999) vol. **1**, pp. 111-118.
34. Shen D., Jale S.R. , Bülow M., and Ojo A.F., *Stud. Surf. Sci. Catalysis* **125** (1999) 667-674.
35. Shen D., Bülow M. , Jale S.R., Fitch F.R., and Ojo A.F., *Micropor. Mesopor. Mater.* **48** (2001) 211-217.
36. Shen D., Bülow M., Siperstein F., Engelhard M., and Myers A.L., *Adsorption* **6** (2000) 275-286.
37. Shen D., Engelhard M., Siperstein F., Myers A.L., and Bülow M., in *Adsorption Science and Technology,* ed. by D.D. Do (Word Scientific, Singapore, 2000) pp. 106- 110.
38. Rees L.V.C. and Shen D. , *Stud. Surf. Sci. Catalysis* **137** (2001) 579-631.
39. Fitch, F.R., Bülow, M., Ojo, A.F., US Patent No. 5,464,467; 1995.
40. Ojo A.F., Fitch F.R., and Bülow M., US Patent No. 5,531,808; 1996.
41. Hill T.L., *J. Phys. Chem.* **17** (1949) 520-533.
42. Fomkin A.A., Serpinsky V.V., and Fiedler K., *Izv. Akad. Nauk USSR, Ser. Chim.* 1982, 1207-1214.
43. Fomkin A.A., in *6th Conf. Theoret. Problems Adsorption: Theses of Lectures* (Russ.), ed. by V.V. Serpinsky and K.O. Murdmaa (Acad. Sci. USSR, Moscow, 1985) pp. 9-17.

44. Bräuer P., Heuchel M., Kalies G., and Messow U., *Chem. Technik (Leipzig)* **50** (1998) 57-64.
45. Huang Y.Y., Bensen J.E., and Boudart M., *IE&C Fundamentals* **8** (1969) 346-353.
46. Stroud H.J.F.and Parsonage N.G., *Adv. Chem. Ser.* **102** (1971) 138-142 (*Proc. 2d Internat. Conf. Molecular Sieve Zeolites*, Worcester, 1970) (ACS, Washington, 1970).
47. Sichhart K.-H., Kölsch P., and Schirmer W., *Adv. Chem. Ser.* **102** (1971) 132-137 (*Proc. 2d Internat. Conf. Molecular Sieve Zeolites*, Worcester, 1970) (ACS, Washington, 1970).
48. Bülow M. and Schirmer W., *Z. physik. Chem. (Leipzig)* **253** (1973) 130-135. 49. Gusev V.Yu., *Adsorption Calorimetry in Broad Ranges of Pressures and Temperatures* (Russ.), Dissertation: Candidate Phys.-Math. Sci. (Inst. Phys. Chem., Acad. Sci. USSR, Moscow, 1991).
49. Gusev V.Yu., *Adsorption Calorimetry in Broad Ranges of Pressures and Temperatures*(Russ.), Dissertation: Candidate Phys.-Math. Sci. (Inst. Phys. Chem., Acad. Sci. USSR, Moscow, 1991).
50. Fomkin A.A., *Physical Adsorption of Gases, Vapors and Liquids by Microporous Adsorbents at High Pressures* (Russ.), Dissertation: Dr. Sci. (Inst. Phys. Chem., Russian Acad. Sci., Moscow, 1993).
51. Kiselev A.V., *Intermolecular Interactions in Adsorption and Chromatography* uss.) (Vys'shaja Shkola, Moscow, 1986) Chapter 13.
52. Karapetjanz M.Ch., *Chemical Thermodynamics* (Russ.) (Gos. Chim. Isdat., Moscow, 1953) p. 57, p.193.
53. Bülow M., *Adsorption*, to be submited.
54. Bering B.P., Myers A.L., and Serpinsky V.V., *Dokl. Akad. Nauk USSR* **193** (1970) 119-122.
55. Myers A.L. and Prausnitz J.M., *AIChE J.* **11** (1965) 121-127.
56. Atkins P.W., *Physical Chemistry*, University Press, Oxford 1978, p. 112.
57. Bülow M. and Werner U., *Z. physik. Chem. (Leipzig)* **259** (1978) 732-736.
58. Allen M.P.and Tildesley D.J., *Computer Simulation of Liquids* (University Press, Oxford, 1956).
59. Watanabe K., Austin N., and Stapleton M.R., *Mol. Simul.* **15** (1995) 197-221. 60. Jale S.R., Bülow M., Fitch F.R., Perelman N., and Shen, D., *J. Phys. Chem. B* **104** (2000) 5272-5280.
60. Jale S.R., Bülow M., Fitch F.R., Perelman N., and Shen, D., *J. Phys. Chem. B* **104**(2000) 5272-5280.
61. Bülow M., Shen D., and Jale S.R., in *Adsorption Science and Technology* (Proc. 3d Pacific Basin Conf. Adsorption Science and Technology, Kyongju, Korea, May 25-29, 2003) ed. by Chang-Ha Lee (World Scientific, Singapore, London, 2003) pp. 114-120.

62. The Matheson Company, *Matheson Gas Data Book*, 4th Edition (East Rutherford, NJ, 1966) p. 83 (CO_2), p. 387 (N_2O).
63. Shen D.M., Huggahalli M., Bülow M., Jale S.R., and Kumar R., US Patent No. 6,391,092; 2002.
64. Bülow M., *Adsorption*, submitted.
65. Garrett, M.E. and Shervington, E.A., Intern. Patent Publ. No. WO 97/47932, Appl. No. PCT/GB97/00045, December 18, 1997.
66. Lopatkin A.A., *Theoretical Fundamentals of Physical Adsorption* (Russ.) (Moscow University Publ., Moscow, 1983) p. 118.
67. Bülow M., Dougill B., Sajik B. and Parkyns N., US Patent No. 6,006,797; 1999.
68. Toufar, H., Toufar, S., Maher, P.K., Ojo, A.F., Fitch, F.R., and Bülow, M., US Patent No. 5,916,836; 1999.
69. Brandt, A., Unger, B., Tschritter, H., Bülow, M., Fitch, F.R., Ojo, A.F., US Patent No. 6,407,025; 2002.
70. Chao, C.C., US Patent No. 4,859,217; 1989.
71. Bülow M., Jale S.R., Ojo A.F., Fitch F.R., and Shen D., *Stud. Surf. Sci. Catal.* **154B** (2004) 1961-1970 (*Proc. 14th Intern. Zeolite Conference* ed. by E. van Steen, M. Claeys and L.H. Callanan) (Elsevier, Amsterdam, 2004).
72. Bakaev V.A. and Steele W.A, in *Fundamentals of Adsorption*, ed by M.D. LeVan (Kluwer Academic Publ., Boston, Dordrecht, London, 1996) pp. 83-90.
73. (a) Siperstein F.R and Myers A.L., *AIChE Journal* **47** (2001) 1141-1159.
(b) Myers A.L., *Adsorption* **11** (2005) 37-42.
74. Mishin I.V., Brueva T.R. and Kapustin G.I., *Adsorption* **11** (2005) 415-424.

SUPERCRITICAL ADSORPTION MECHANISM AND ITS IMPACT TO APPLICATION STUDIES

L. ZHOU, Y. SUN, W. SU AND Y. P. ZHOU

High Pressure Adsorption Laboratory, School of Chemical Engineering & Technology

Tianjin University, Tianjin 300072, China.

E-mail: zhouli-tju@eyou.com

Hydrogen storage and methane capture receive the worldwide attention due to their importance in sustainable energy and environment protection. Adsorption provides an efficient way to compress gases, therefore, has been applied for the development of hydrogen storage technology. It also provides an efficient way to separate gas mixtures, therefore, is being studied for the capture of methane from its mixture with air in order to avoid methane emission. However, both hydrogen and methane are supercritical gases at the temperature of engineering interest and follow a different mechanism of adsorption compared to that of sub-critical gases. The present work shows why only monolayer coverage mechanism functions at above-critical temperatures. Pros and cons to this point of view are presented. This understanding of the adsorption mechanism is essential for the research of hydrogen storage since the mechanism claims that any storage method based on adsorption will not satisfy the commercial requirement for hydrogen storage no matter how novel the material is. On the other hand, understanding the adsorption mechanism may help to follow a successful route in the research. Development of an efficient adsorbent for methane capture from its mixture with air is such an example.

1. Introduction

Energy source and environment protection are problems of common concern. Adsorption of gases is the basis of quite a few technologies that are of great potential for solving various problems; therefore, it has attracted a great deal of research interest recently. Adsorption yields an efficient technology usually for gas or gas mixtures of small molecular weights. Hydrogen and methane are two gases of special importance for both energy source and environment protection. Hydrogen is considered a renewable and sustainable energy carrier, and many projects are being carried out worldwide to develop hydrogen-fueled vehicles. However, an on-board storage of hydrogen is still the major technical barrier on the way to utilize hydrogen energy, although many efforts have been dedicated to the solution of the problem. Methane is also an important gas not only because its abundance on the earth but also due to its greenhouse effect, which is much stronger than that of carbon dioxide. The abundance of methane considerably increased since the discovery of flammable ice. A lot of methane is stored in coal beds either, but most of them are blown off into atmosphere as the effluent of coalmines. A lot of clean fuel is lost this way and the environment is damaged either. Therefore, how to capture methane is very important for both the reduction of greenhouse effect and the utilization of clean fuels. A huge amount of application studies have been carried out, however, to find out the solution of these problems would still be a serious challenge if adsorption mechanism of these gases remains unclear. The critical temperature of gases with small molecular weights is low. For example, the critical temperature of hydrogen is 33 K, and that of methane is 190.6 K. Therefore, these gases are supercritical and incondensable at the temperatures of engineering interest. They must follow a different adsorption mechanism than that of condensable gases.

2. Adsorption mechanism of condensable gases

A fundamental law of physics claims that fluid at a temperature higher than the critical one is incondensable or cannot be liquefied no matter how high pressure is applied, although it is condensable vapor or can be liquefied at sub-critical temperatures. All experimental data available today show that the adsorption isotherms of vapors can be classified into six types depending on the structural (geometrical) properties of adsorbents [1]. The six type isotherms have a common feature, i.e., the amount adsorbed increases unimodally with pressure. The mechanism of vapor adsorption might be monomolecular surface coverage, multimolecular surface coverage, volume filling or capillary condensation. All the mechanisms rely on the possibility of condensation under the adsorption condition. This kind of adsorption phenomena can be well explained by the existing adsorption theories, and these theories are utilized to characterize adsorbents on the basis of experimental adsorption isotherms.

3. Adsorption mechanism of incondensable gases

Since gas cannot be liquefied at temperatures higher than the critical one, the adsorbed gas cannot be liquid-like either no matter how strong the interaction between the gas molecules and the lattice atoms of solid surface is. Therefore, all adsorption mechanisms relying on condensation including volume filling, multimolecular coverage and capillary condensation will not function at above-critical temperatures. What can and really occur is merely monomolecular surface coverage.

There are multiple arguments supporting the claim of monolayer adsorption mechanism at above-critical temperatures.

3.3. *The unique form of adsorption isotherms*

So far, only one type of supercritical adsorption isotherms has been experimentally observed no matter how different the adsorbents are. The common feature of supercritical adsorption isotherms is the existence of an isotherm maximum. The isotherm looks like type-I before the maximum and decreases after it. Zero, even negative amount adsorbed was experimentally recorded [2]. It is well known that the isotherm shape is governed by the underlying adsorption mechanism; therefore, the unique isotherm shape must reflect the unique adsorption mechanism.

3.2. *Implication arising from the BET theory of adsorption*

The well-known BET theory of adsorption is still the basis of evaluating the specific surface area of porous solids [3]. It claims that the first molecular layer is fixed on the solid surface due to the interaction between gas and solid. More gas molecules may be adsorbed above the first adsorbed layer due to the interaction among the adsorbate molecules forming the second and subsequent layers. The interaction energy between the first layer adsorbates and the surface atoms differs from that among the adsorbates in the second and subsequent layers. This difference must be reflected in the heat of adsorption of different layers. The experiment for nitrogen adsorption on carbon black [4] showed that the heat of adsorption for the first layer is 11 to 12 kJ/mol (0.11 to 0.12 eV) and it drops to 5.56 kJ/mol (0.058 eV) in the subsequent layers. The latter is quite the same as the latent heat of condensation. Obviously, the second and subsequent layers cannot exist at above-critical temperatures due to the incondensability of gases.

3.3. *Evidence arising from hydrogen adsorption experiments*

Carbon materials are considered promising for hydrogen storage and a vast variety of experiments have been performed for this purpose. The volume of the adsorbed hydrogen evaluated on the basis of storage capacity for a microporous activated carbon is only 0.4 and 0.24 cm^3/g for powder and pellets, respectively, as shown in Fig. 1. This volume is considerably less than the pore volume of

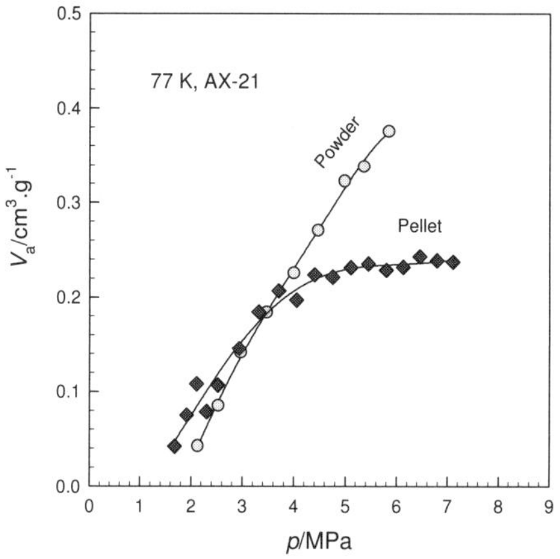

Figure 1. Volume of the adsorbed phase evaluated on the basis of H_2 storage capacity [5].

the carbon, 1.3 cm^3/g [5]. Therefore, volume-filling mechanism did not function. Ströbel et al [6] measured the hydrogen uptake capacity for a series of carbon materials with a high-pressure microbalance at 12.5 MPa and 296 K. The BET surface area of the tested materials ranged from 100 to 3300 m^2/g. Hydrogen uptake capacity was found to be proportional to the specific surface area of adsorbents as described by Equation 1.

$$\text{wt \%} = 0.0005.S\ [\text{m}^2.\text{g}^{-1}] \qquad (1)$$

Nijkamp and coworkers [7] also reported the linear relationship between hydrogen adsorption and the specific surface area of adsorbents on the basis of hydrogen adsorption capacity measured for many carbon materials at 77 K. This relationship exists only when adsorption of hydrogen is monolayer. The author's lab collected adsorption isotherms of hydrogen isotopes on 21 micro- and mesoporous molecular sieves made of different materials [8]. The amount adsorbed at 77 K and 0.1 MPa was plotted against the specific surface area of adsorbents as shown for H_2 and D_2 in Figure 2. Linearity of the dependence is clearly shown for all adsorbents no matter carbonaceous or not. Furthermore, the slopes of the linearity are remarkably different in the microporous section (including 15 adsorbents) and the mesoporous section (including 6 adsorbents), and a little difference between H_2 and D_2 is observed in each section either. The fact that adsorption capacities of adsorbents made of different materials locate on unique linear plot is a convincing proof of the claim that hydrogen adsorption

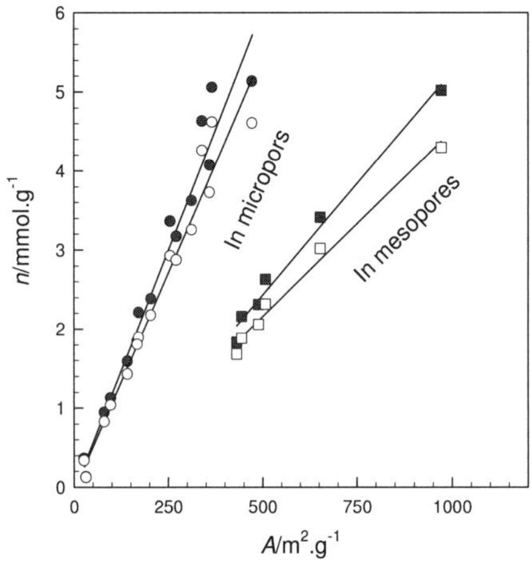

Figure 2. Dependence of adsorption amount on specific surface area [8]. Light marks: H_2; Dark marks: D_2.

can only be monolayer coverage on the adsorbent surface and the surface property is not important for the adsorption capacity.

3.4. *Evidence arising from modeling adsorption isotherm*

Numerous efforts have been made to explain the abnormal behavior of supercritical adsorption isotherms and several theories were proposed. Overheated liquid [9] or quasi-liquid [10] conceptions were used to model the supercritical adsorption isotherms on the basis of the theory available for vapors. However, isotherms with maximum cannot be described in this way. The model based on the Ono-Kondo equation [11] was able to predict an isotherm with maximum, but its parameters were found to be unrealistic from the physical viewpoint [12]. Models based on the equation of state [13] and density functional theory [14] can satisfactorily describe the experimental adsorption isotherms. However, the number of parameters in such models is much larger than 3, the usual number of parameters in conventional isotherm equations. In fact, the multiple model parameters cannot provide the required information about adsorbents regarding their specific surface area, pore-volume and pore size distribution as it was usually done with conventional isotherm models.

The authors explained the abnormal behavior of supercritical adsorption isotherms on the basis of the Gibbs definition of adsorption [15]. The definition shown in Eq. 2 applies for adsorption under any condition.

$$n = n^s - \rho_g V_a = V_a(\rho_a - \rho_g) \qquad (2)$$

Where V_a is the volume of the adsorbed phase, ρ_a and ρ_g are the densities of the adsorbed and gas phase, respectively. In Eq. 2, n is a density-excess quantity and is named as the surface excess adsorption, and n^s is the total quantity of adsorbate in the adsorbed phase and is named as absolute adsorption. The abnormal behavior of isotherms is originated in the difference between the excess quantity and the absolute quantity. This difference is negligible for vapor adsorption since the adsorption pressure cannot be higher than the saturation pressure, at which condensation occurs and adsorption ends. Therefore, the density of the vapor phase cannot be high. On the other hand, the state of the adsorbed adsorbate is quite close to liquid; therefore, the difference between the two phase densities is so large that the second term of the right hand side of Eq. 2 is negligible and

$$(\rho_a - \rho_g) \approx \rho_a \Rightarrow n \approx n^s$$

It is clear that the adsorption isotherm of vapors is indeed the isotherm of absolute adsorption. Since all isotherm models were initially developed for absolute adsorption, they can fit the experimental isotherms. However, there is not a satration pressure at above-critical temperatures, and the gas density, ρ_g, always increases with the increasing pressure. The density of the adsorbed phase, ρ_a, on the other hand, is limited by the smallest clearance between molecules and the limited strength of inter-molecular interactions. Therefore, the difference between the two phase densities, $(\rho_a - \rho_g)$, becomes smaller and smaller with the increasing adsorption pressure, until the isotherm maximum appears; after which the recorded amount adsorbed decreases and even becomes zero or negative. Obviously, direct application of the conventional isotherm models cannot describe the experimental adsorption isotherms at above-critical temperatures due to the increasing difference between the absolute and the excess adsorption. Therefore, this difference must be evaluated for the proper dscription of supercritical adsorption. However, the absolute quantity of adsorption cannot experimentally be determined under commonly used conditions, and the determination of the absolute adsorption quantity on the basis of experimentally collected excess isotherms has been considered an essential problem or a challenge in the study of supercritical adsorption [16, 17].

On the basis of equality of the excess and the absolute quantities of adsorption for the condition of dilute surface concentration, the authors proposed a method to predict the absolute adsorption on the basis of the experimental excess adsorption data. As a consequence, the difference between the excess and the absolute adsorption was evaluated [18, 19]. The second term in the right hand side of Eq. 2 would not contain any unknowns, and any isotherm equation available for monolayer adsorption would be able to apply for n^s in the equation [20]. The traditional adsorption theory was thus extended to the area of supercritical temperatures. Applying an isotherm equation tailored for monolayer adsorption mechanism, Eq. 2 satisfactorily describes the experimental high-pressure adsorption isotherms available till today as shown in Figures 3-6 as examples [21-24].

3.5. *Direct evidence of FTIR measurements*

To know how does the adsorption mechanism change following the temperature increase from sub-critical to supercritical region, the author's lab collected CO_2 isotherms on activated carbon at different temperatures, and the average number of molecular layers in the adsorbed phase was calculated [23]. While the number is 1.20 at 307 K, it reduces to 1.0 and less at 323 K and higher temperatures. Although 307 K is higher than the critical temperature (304.2 K), it is still in the critical zone; therefore, multilayer adsorption is possible to occur at some cites. However, as the temperature increases, multilayer adsorption is never observed. This result was further proved by the in situ FTIR spectroscopy for the near-critical CO_2 in mesoporous silica [25]. This study tells whether multilayer or monolayer adsorption really occurred on the surface of adsorbent, and its result is in agreement with ours.

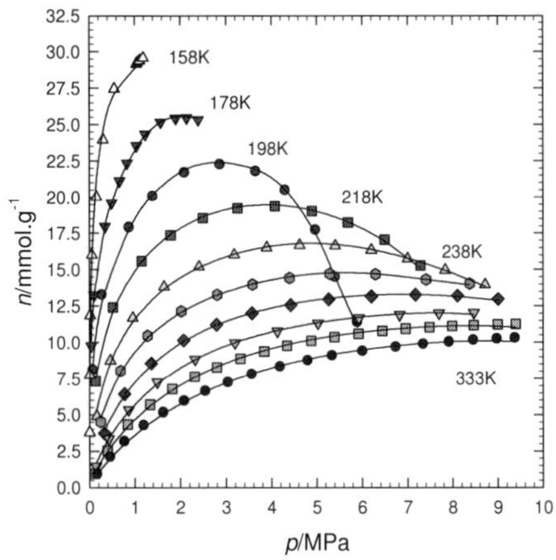

Figure 3. Adsorption isotherms of CH_4 on activated carbon spanning the critical temperature [21]. Dots: data; Curves: model

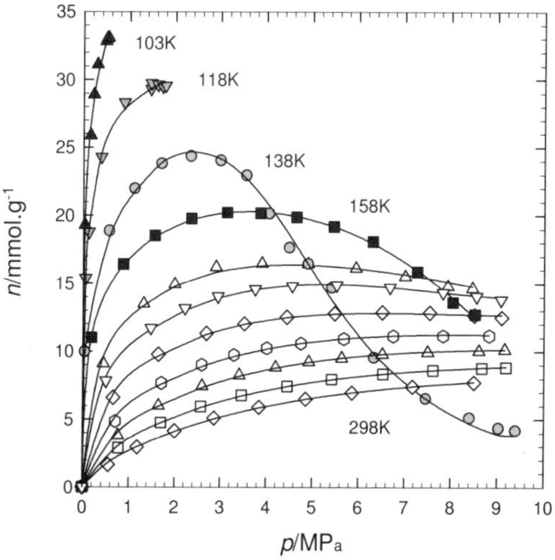

Figure 4. Experimental excess adsorption isotherms of N_2 on activated carbon. Dots: data; Curves: model [22]

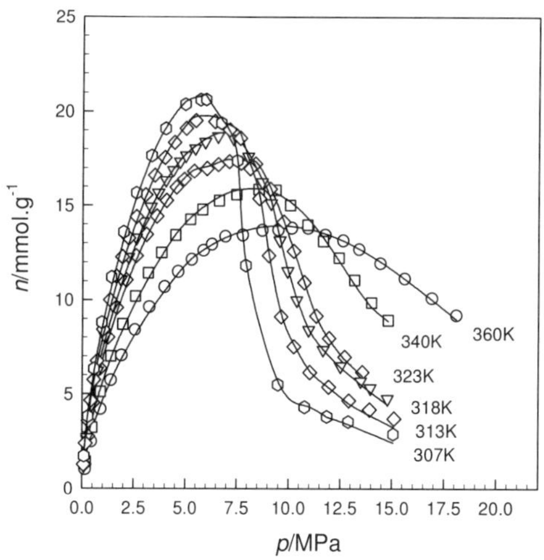

Figure 5. Adsorption isotherms for the supercritical region [23]. Dots: experimental; Curves: predicted by model

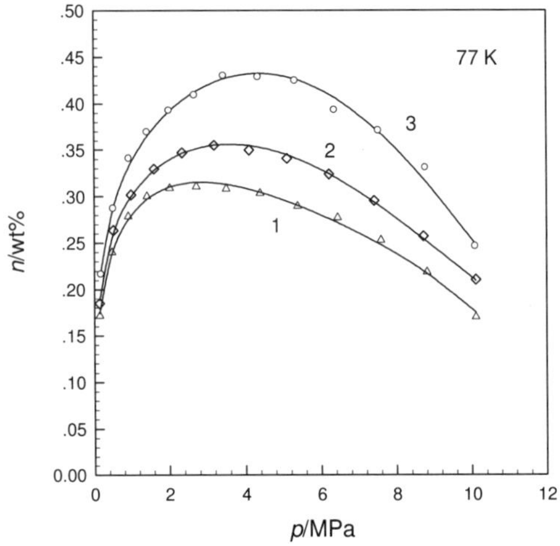

Figure 6. Adsorption isotherms of H2 on MWNT sample. 1: powder before heat treatment; 2: powder after heat treatment; 3: pellets [24]. Dots: experimental; Curves: predicted by model

4. Disputation to the monolayer mechanism

A disputation to the monolayer mechanism claims that gas molecules confined in a space of nano-dimension, such as inside carbon nanotubes, must receive an ultra ordinary action applied by the surrounding walls and the liquid state might be assumed. However, there is not any experimental or molecular simulation proofs to support the claim. According to a molecular dynamics simulation [26], a hydrogen atom with dynamic moment 20 eV was transplanted through the wall into a tube of diameter 0.683 nm composing of 150 carbon atoms. It was found that hydrogen atoms were recombined to form molecules and arranged concentrically inside the tube. Pressure inside the tube reached to 350 thousand bar when the implanted hydrogen atoms were 90 (5 wt %). No condensation was shown even at such high pressure.

Another disputation to the monolayer mechanism comes from the fact that the density of incondensable gas keeps increasing and the molecules tend to settle down orderly above the solid surface, and the ordered multiple layer settlement was attributed to adsorption and, as such, the monolayer mechanism no longer functions. To elucidate why the multiple layer settlement in this case cannot be considered adsorption, one is referred to the fundamental observation and definition of adsorption. Adsorption is a function of pressure, but only for a definite limit, i.e., there is an upper limit of adsorption in any cases. The upper limit is the saturation pressure below the critical temperature. The upper limit still exists for supercritical adsorption, although the saturation pressure disappears [27]. As a fact, adsorption is a phenomenon due to internal forces, i.e., the interaction between molecules/atoms, therefore, any changes in the adsorbed phase caused by an external force cannot be attributed to the phenomenon of adsorption. The upper limit for supercritical adsorption is determined by the balance between the interactions of internal and external forces. As shown in Fig. 7, supercritical adsorption isotherms show a linear section after the maximum if the abscissa is expressed in gas phase density [28]. The volume of the adsorbed phase, V_a, and the total adsorbate quantity in the adsorbed phase, n^s, must be constant if the relation between n and ρ_g is linear according to Eq. 2. It states that the adsorbed phase cannot admit any more molecules to enter. Therefore, adsorption is indeed ended at the beginning of the linear section. The external force may be comparably large to the internal one for the linear range of gas phase density, and finally overtakes the latter and results disturbance in the adsorbed phase at the upper bound of the linear section, and adsorption ends there. It is argued that the gas phase density that enforces the gas molecules to be settled down orderly must be much higher than that when adsorption ends, otherwise the linear section of the adsorption isotherm will not maintain. In fact, the recorded isotherm continues after the linear section, which is really caused by the ever-increasing external force and nolonger belongs to adsorption.

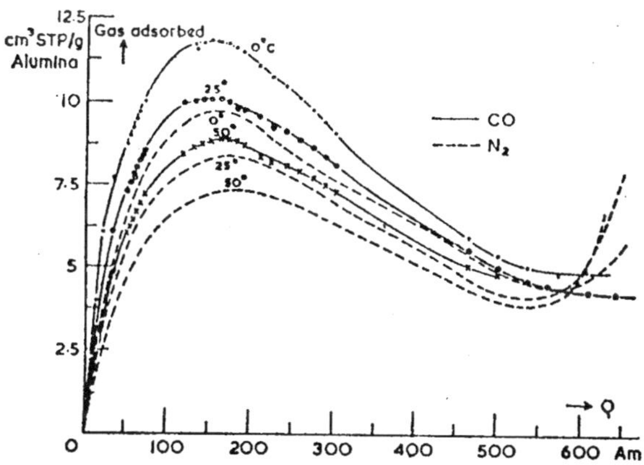

Figure 7. Typical supercritical adsorption isotherms [28]

5. Impact to the research of hydrogen storage

According to the monolayer mechanism of adsorption, hydrogen uptake capacity of any material is limited by the specific surface area of the material should the temperature is remarkably higher than 33 K. Other feature or property of the material will not exert an essential effect on the storage capacity. Carbon nanotubes are not suitable for hydrogen storage due to its small surface area. This adsorption mechanism applies certainly for MOF (metal organic frameworks) material either. Although the state of adsorbed hydrogen may change with pressure [29], physical adsorption dominates the storage since the magnitude of adsorption heat is only 4~9 kJ/mol and the amount adsorbed change inversely with temperature [30]. In addition, the isotherms also show a maximum. Therefore, adsorption of hydrogen on MOF also follows the general rules of supercritical adsorption. There is not much difference in the specific surface area between superactivated carbon and MOF (whose extremely high specific surface area is only claimed by molecular simulation, yet opposed by experimental measurement), and there is not much difference in the hydrogen storage capacities between them either. The storage capacity at ambient temperature is considerably lower than that at low temperatures. Therefore, hydrogen storage based on physical adsorption cannot have as high a storage capacity as set up by motor vehicles producer. Instead of storing hydrogen at ambient temperature, cryogenic storage on superactivated carbon provides a relatively high capacity with a competitive cost [31]. Storage based on chemical adsorption is not suitable for on-board storage either. Chemical adsorption can only follow monolayer mechanism, and it occurs usually at elevated temperatures, which is not preferred from the cost point of view.

6. Impact to the research of methane capture

Methane capture is especially important for coal mining. A huge quantity of methane was blown off into the atmosphere provided methane content is not high enough to be used as fuels, and a great portion of greenhouse effect is contributed by methane this way. Explosion danger exists if the content of methane is in the range of 3-15%. Capture of methane from the coalmine exhaust is, therefore, very important. To practice the capture, an efficient separation between the key components, methane and nitrogen, must be realized. Pressure swing adsorption (PSA) is known to be a simple yet cost-competitive separation technology for mixtures composed of small molecules. However, conventional adsorbents are not efficient for the separation and searching for an efficient adsorbent for the separation between methane and nitrogen remains a challenge

[32]. Adsorptive separation is based on the difference of mixture components in the equilibrium adsorption, rate of adsorption or shape and/or size. The size and molecular weight of the two gases are quite close, and their physical or chemical property is also similar, therefore, the difference in the equilibrium adsorption must be somehow enlarged. Enlightened by the monolayer adsorption mechanism, the author's lab successfully enlarged the separation coefficient for several times [33]. As is shown in Fig. 8, the separation coefficient correlates with the specific surface area of adsorbents linearly. Recently, the feasibility of the PSA separation was further proved by a continuous run on a two-column process in the authors' laboratory. Its practical application in the future will certainly have an important consequence.

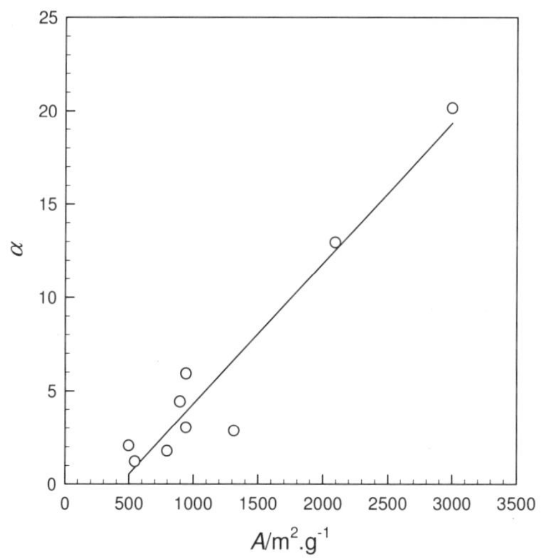

Figure 8. Correlation between the separation coefficient for CH_4/N_2 and the specific surface area of adsorbents

7. Conclusion

Adsorption of hydrogen and methane has been widely studied from the viewpoint of storage and separation. It is important to be aware of that the monolayer adsorption mechanism functions in either physical (at above-critical temperatures) or chemical adsorption. Any effort to enhance hydrogen storage using solid material can hardly reach the commercial goal as long as this enhancement is based on adsorption. On the other hand, an efficient adsorbent for methane capture is successfully developed under the guidance of the monolayer adsorption mechanism.

Acknowledgements

The authors thank the National Natural Science Foundation of China for its consecutive support for the research (under grant number 59543011, 29676031, 29936100 and 20336020).

References

1. IUPAC Commission on Colloid and Surface Chemistry Including Catalysis, *Pure Appl. Chem.*, **57** (1985) pp.603-619.
2. L. Zhou, in J. Toth (ed.), *Adsorption: Theory, Modeling & Analysis*. 2002, Marcel Dekker, New York, pp. 211-250.
3. S. Brunauer, P. H. Emmett, E. Teller. Adsorption of gases in multi-molecular layers. *J. Am. Chem. Soc.***60** (1938) 309-.
4. B. A. Beebe, J. Biscoe, W. R. Smith, C. B. Wendell, Heats of Adsorption on Carbon Black. I. *J. Am. Chem. Soc.* **69** (1947) pp.95-101.
5. L. Zhou, Y. Sun, Y. P. Zhou, Storage of Hydrogen on Carbon Materials: Experiments and Analyses. *Chem Eng Commun*. **193** (2006) pp.564-579.
6. R. Ströbel, L. Jörisen, T. Schliermann, V. Trapp, W. Schütz, K. Bohmhammel, G. Wolf, J. Garche. Hydrogen adsorption on carbon materials. *J. Power Source*. **84** (1999) pp.221-224.
7. M. G. Nijkamp, J. E. M. J. Raaymakers, A. J. van Dillen, K. P. de Jong. Hydrogen storage using physisorption - materials demands. *Appl. Phys. A*. **72** (2001) pp.619-623.
8. X. Z. Chu, Y. P. Zhou, W. Su, Y. Sun, L. Zhou. Adsorption of Hydrogen Isotopes on Micro- and Mesoporous Adsorbents with Orderly Structure. *J Phys Chem B*. **110**(2006) pp.22596-22600.
9. S. Ozawa, S. Kusumi, Y. Ogino. Physical adsorption of gases at high pressure, IV. An improvement of the Dubinin-Astakhov adsorption equation. *J. Colloid & Interface Sci*. **56** (1976) pp.83-91.
10. K. Kaneko, K. Shimizu, T. Suzuki. Intrapore field-dependent micropore filling of supercritical N_2 in slit-shaped micropores. *J. Chem. Phys.* **97** (1992) 8705-8711.
11. G. Aranovich, M. Donahue. Adsorption of Supercritical Fluids. *J. Colloid & Interface Sci*. **180** (1996) 537-541.
12. P. Bénard, R. Chahine. Modeling of High-Pressure Adsorption Isotherms above the Critical Temperature on Microporous Adsorbents: Application to Methane. *Langmuir*. **13** (1997) 808-813.
13. E. A. Ustinov, D. D. Do, A. Herbst, R. Staudt, P. Harting. Modeling of

gas adsorption equilibrium over a wide range of pressure: A thermodynamic approach based on equation of state. *J Colloid & Interface Sci.* **250** (2002) 49-62.

14. V. Neimark, P. I. Ravikovitch. Calibration of adsorption theories. in F. Meunier (ed.), *Proceedings FOA6*, 1998, Elsevier, Paris, 159-164.
15. D. H. Everett, *Manual of symbols and terminology for physicochemical quantities and units. Appendix II. Part I*, 1971, Butterworth, London.
16. M. M. K. Salem, P. Braeuer, M. V. Szombathely, M. Heuchel, P. Harting, K. Quitzsch, M. Jaroniec. Thermodynamics of high-pressure adsorption of argon, nitrogen, and methane on microporous adsorbents. *Langmuir.* **14** (1998) 3376-3389.
17. K. Murata, K. Kaneko. Nano-range interfacial layer upon high pressure adsorption of supercritical adsorption of supercritical gases. *Chem Phys Lett.* **321** (2000) 342-348.
18. L. Zhou, Y. P. Zhou. Linearization of Adsorption Isotherms for High Pressure Applications. *Chem Eng Sci.* **53** (1998) 2531-2536.
19. L. Zhou, Y. P. Zhou. A Mathematical Method for the Determination of Absolute Adsorption from Experimental Isotherms of Supercritical Gases. *Chinese J Chem Eng.* **9** (2001) 110-115.
20. L. Zhou, Y. P. Zhou, M. Li, P. Chen, Y. Wang. Experimental and Modeling Study of the Adsorption of Supercritical Methane on a High Surface Carbon. *Langmuir.* **16** (2000) 5955-5959.
21. L. Zhou, J. S. Zhang, Y. P. Zhou. A Simple Isotherm Equation for Modeling the Adsorption Equilibria on Porous Solids over Wide Range Temperatures. *Langmuir.* **17** (2001) 5503-5507.
22. L. Zhou, Y. P. Zhou, S. P. Bai, C. Z. Lü, B. Yang. Determination of the Adsorbed Phase Volume and Its Application in Isotherm Modeling for the Adsorption of Supercritical Nitrogen on Activated Carbon. *J. Colloid & Interface Sci.* **239** (2001) 33-38.
23. L. Zhou, S. P. Bai, W. Su, J. Yang, Y. P. Zhou. Comparative Study of the Excess versus Absolute Adsorption of CO_2 on Superactivated Carbon for the Near-Critical Region. *Langmuir.* **19** (2003) 2683-2690.
24. Y. P. Zhou, K. Feng, Y. Sun, L. Zhou. Adsorption of Hydrogen on Multiwalled Carbon Nanotubes at 77 K. *Chem. Phys. Lett.* **380** (2003) 526-529.
25. M. S. Schneider, J. D. Grunwaldt, A. Baiker. Near-Critical CO_2 in Mesoporous Silica Studied by In Situ FTIR Spectroscopy. *Langmuir.* **20** (2004) 2890-2899.
26. Y. Ma, Y. Xia, M. Zhao, R. Wang, L. Mei. Effective hydrogen storage in single-wall carbon nanotubes. *Phys. Rev. B.* **63** (2001) 115422.
27. L. Zhou, Y. P. Zhou, S. P. Bai, B. Yang. Studies on the Transition Behavior of Physical Adsorption from the Sub- to the Supercritical

Region: Experiments on Silica Gel. *J. Colloid & Interface Sci*. **253** (2002) 9-15.

28. P. G. Menon. Adsorption at High Pressures. *Chem. Rev*. **68** (1968) 277-294.
29. D. Y. Siberio-Pèrez, O. M. Yaghi, A. J. Matzger. Adsorption of Hydrogen and Methane in Metal-Organic Frameworks. 2005 AIChE Annual Meeting, 440c, Cincinnati.
30. G. Garberoglio, A. Skouliddas, K. Johnson. Mechanism of Hydrogen Adsorption in Metal Organic Frameworks. 2005 AIChE Annual Meeting, 440d, Cincinnati.
31. L. Zhou, Y. P. Zhou, Y. Sun. Enhanced Storage of Hydrogen at the Temperature of Liquid Nitrogen. *Int. J. Hydrogen Energy*. **29** (2004) 319-322.
32. D. M. Ruthven. Past Progress and Future Challenges in Adsorption Research. *Ind Eng Chem Res*. **39** (2000) 2127-2131.
33. L. Zhou, W. C. Guo, Y. P. Zhou. A Feasibility Study of Separating CH_4/N_2 by Adsorption. *Chinese J. Chem. Eng*. **10** (2002) 558-561.

Part B: Fundamental

STRUCTURAL MODELING OF POROUS CARBONS USING A HYBRID REVERSE MONTE CARLO METHOD

S. K. JAIN AND R. J-M. PELLENQ

CNRS, Campus de Luminy, Case 913 13288 Marseille cedex 09, France.
E-mail : skjain2@unity.ncsu.edu, pellenq@crmcn.univ-mrs.fr

K. E. GUBBINS

Center for High Performance Simulation and Department of Chemical and Biomolecular Engineering, North Carolina State University at Raleigh, Box 7905, Raleigh, NC 27695-7905, U.S.A.
E-mail: keg@ncsu.edu

We present molecular models for 3 saccharose based carbons of different densities obtained using a Reverse Monte Carlo (RMC) protocol which incorporates an energy constraint. The radial distribution functions of the simulated models are in good agreement with experiment. Moreover, 3 and 4 member carbon rings, reported in the literature for many modeling studies of carbon, are absent or extremely rare in our final structural models. These small member rings are high energy structures and are believed to be an artifact of the usual RMC method. The presence of the energy penalty term in our simulation protocol penalizes the formation of these structures. Using a ring connectivity analysis method that we developed, we find that these atomistic models of carbons are made up of defective graphene segments twisted in a complex way. These graphene segments are largely made up of 6 carbon member rings, but also contain some 5 and 7 carbon member rings. We also found that in addition to the graphene segments there are some carbon chains which do not belong to any graphene segments. To characterize our models, we calculated the geometric pore size distribution and also simulated the adsorption of argon at 77.4 K in the models using GCMC simulations. The adsorption isotherm obtained for all three models are representative of microporous carbons.

1. Introduction

Porous carbons are disordered materials with heterogeneous pore structures. These materials are usually modeled using the slit pore model, in which the material is assumed to be made up of independent and unconnected pores. However this model fails to account for the complicated pore geometry and also the pore connectivity present in the real porous carbons. In recent times, reconstruction methods have been popular to develop realistic molecular models of these materials. In this approach a 3D structural model is built that is consistent with a set of experimental data. Reverse Monte Carlo (RMC) [1] is one such reconstruction method, in which the molecular model is built to match experimental structure factor data from X-ray or neutron diffraction.

RMC is a fitting procedure in which (subject to some constraints) the model is adjusted to best fit g(r) from experiment. In a previous work [2] we studied the stability of the models obtained from a constrained RMC procedure [3] for saccharose - based carbons by relaxing them using two different approaches that realistically describe the interaction between the carbon atoms. We found that the local structure of these models change upon relaxation. Moreover, these models contain some 3 and 4 member rings; these are eliminated upon relaxation.

In a more recent work we presented a method [4], based on Hybrid Reverse Monte Carlo (HRMC), in which the algorithm attempts to simultaneously minimize the error in the radial distribution function and also the total energy of the system. This is achieved by adding an energy penalty term in the original RMC procedure. The presence of the energy term decreases the probability of having unrealistic structures, while simultaneously matching the experimental data. The use of such an energy term in the acceptance probability of the RMC procedure has been used before by Snook and coworkers [5,6] in the study of amorphous carbons.

We use our simulation protocol [4] to develop molecular models for three porous carbons obtained from saccharose, previously used by Pikunic et al. [3] and named CS400, CS1000, and by Jain et al. and named CS1000a [7]. Here 400 and 1000 represent the temperatures at which these materials are carbonized while 'a' indicates subsequent activation in a CO_2 atmosphere. We develop molecular models by considering carbon and hydrogen atoms and neglect the presence of other hetero atoms. The amount of carbon and hydrogen present in the samples is obtained from the composition data [3,4]. The carbon-carbon, carbon-hydrogen and hydrogen-hydrogen interactions are modeled using the Reactive Empirical Bond Order (REBO) potential [8].

2. Hybrid Reverse Monte Carlo

The Reverse Monte Carlo method was initially proposed by McGreevy and Pustzai [1]. The idea is to generate an atomic configuration of a system that matches the structural properties of the real system obtained by experiment. Throughout the simulation the differences between the simulation and experimental structural properties are minimized. The most commonly used structural property in RMC methods is the structure factor, S(q) and the quantity to be minimized is

$$\chi^2 = \sum_{i=1}^{n\,\exp} \left[S_{sim}(q_i) - S_{\exp}(q_i) \right]^2 \tag{1}$$

where S_{sim} is the structure factor for the model material and S_{exp} is the experimental structure factor. After determining S_{sim} for a given atomic configuration, atoms are moved randomly in a Monte Carlo procedure to obtain a new configuration. The probability of acceptance of a new atomic configuration is given by

$$P_{acc} = \min\left[1, \exp\{-\frac{1}{T_\chi}(\chi^2_{new} - \chi^2_{old})\} \right] \tag{2}$$

where T_χ is a weighting parameter.

In our simulation protocol we introduce an energy penalty term in the acceptance criteria. The energy of the system (C-C, C-H and H-H interactions) is calculated using the REBO potential of Brenner [8], which is based on Tersoff's covalent bonding formalism [9],

$$U_{ij} = V_{ij}^R(r_{ij}) + b_{ij} V_{ij}^A(r_{ij}) \tag{3}$$

It has a pair repulsive, V_{ij}^R, a pair attractive, V_{ij}^A, potential term and a bond order term, b_{ij}, which weights the attractive part of the potential with respect to the repulsive part. The bond order term is a many body term, which depends on the local environment of atoms i and j. A variety of chemical effects that affect the strength of the covalent bonding interaction are all accounted for in this term. Coordination numbers, bond angles and conjugation effects all contribute to the strength of a particular bonding interaction in the REBO potential. The REBO potential is a short ranged potential and does not contain any dispersion interactions. The probability of acceptance of the new atomic configuration is given by:

$$P_{acc} = min\left[1, exp\left(-\frac{1}{T_\chi}\left(\left(\chi^2_{new} - \chi^2_{old}\right) + \frac{1}{w}\left(U_{new} - U_{old}\right) \right) \right) \right] \tag{4}$$

where U_{new} and U_{old} are the energies of the new and old configurations respectively, and w is a weighting parameter used to weight the energy term with respect to the structure one.

3. Results

We used the HRMC procedure, described in the previous section, to build molecular models for 3 carbon samples named CS400, CS1000 and CS1000a. A box size of 25 angstrom was used to build the molecular models for all the samples. The density of the samples as obtained from Hg porosimetry [3,7] are: 1.275 g/ml (CS400), 1.584 g/ml (CS1000) and 0.722 g/ml (CS1000a) respectively. The molecular models were developed by considering carbon and hydrogen atoms. All other heteroatoms present were neglected. We show a comparison between the simulated and experimental radial distribution functions for all the three samples in Figure 1.

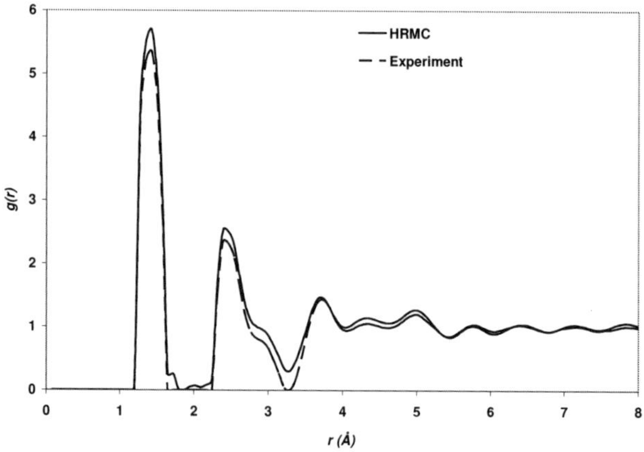

a)

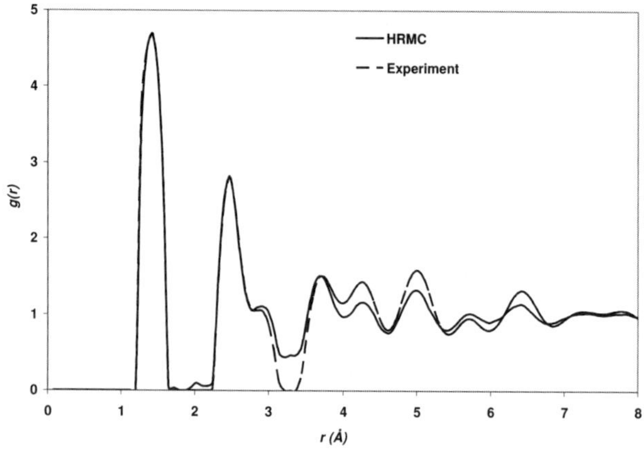

b)

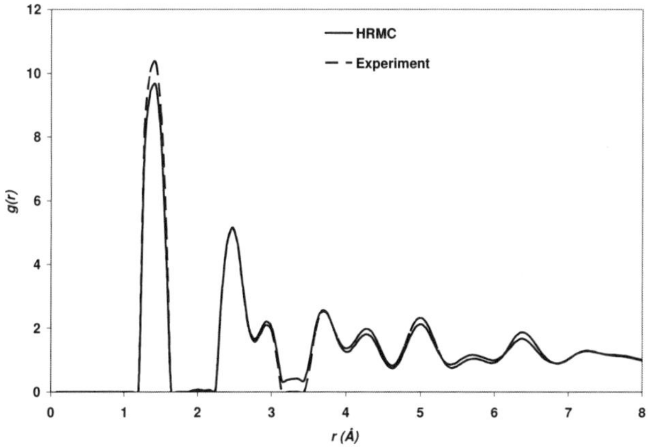

c)

Figure 1. Pair correlation functions obtained from experiment and from the model. (a) CS400, (b) CS1000 and (c) CS1000a

From the above figures we can see that the experimental and simulated radial distribution functions are in good agreement for all the three samples. Upon comparing the pair correlation functions of the three samples it can be seen that CS1000a has more structure as compared to the other two samples, since the peaks are more pronounced and also it has long range correlations.

In atomistic models of amorphous materials, ring statistics provide a measure of medium range order. However, while ring statistics tell us the number of rings of various sizes present in the model, they do not give us any information about the arrangement of rings, e.g. if the rings are clustered and how big is a cluster. In a recent work [10] we presented a method to calculate the ring connectivity, or clustering of rings. We first calculate the rings present in the model using the shortest path criteria of Franzblau [11], and then find the rings that are connected together and group them into clusters. We find clusters containing 5-, 6- and 7- carbon member rings in our models. After isolating the clusters, we found that they resemble defective graphene segments twisted in a complex way. In figure 2 we show snapshots of the molecular models obtained using our simulation protocol. The different color codes represent different graphene segments present in the models.

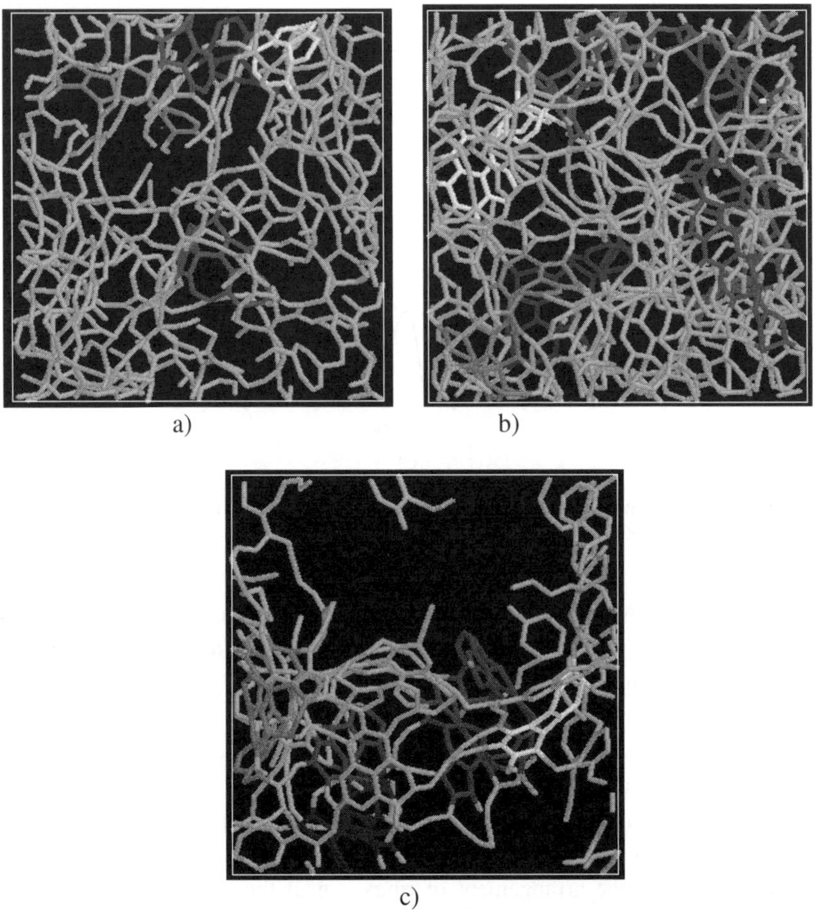

Figure 2. (a) Snapshot of CS400 model obtained from the simulations. The different color code (except grey) represent different graphene segments. (b) the same for CS1000. (c) the same for CS1000a.

Upon analyzing the graphene segments in the resultant models we found that the number and size of the graphene segments (the number of 5, 6 and 7 member carbon rings present in a graphene segment) vary for the three models. Apart from the graphene segments there are many carbon atoms which do not belong to any of the graphene segments and are arranged in a chain like fashion. CS400 is mainly composed of carbon atoms arranged in a chain fashion as can be seen from figure 2(a).

To further characterize our models we calculated the geometric pore size distribution (PSD) using the method of Gelb and Gubbins [12]. The PSDs, as shown in figure 3, reveal that both CS400 and CS1000 contain narrow micropores, whereas CS1000a has a wide PSD with the maximum pore size going to 12 angstrom.

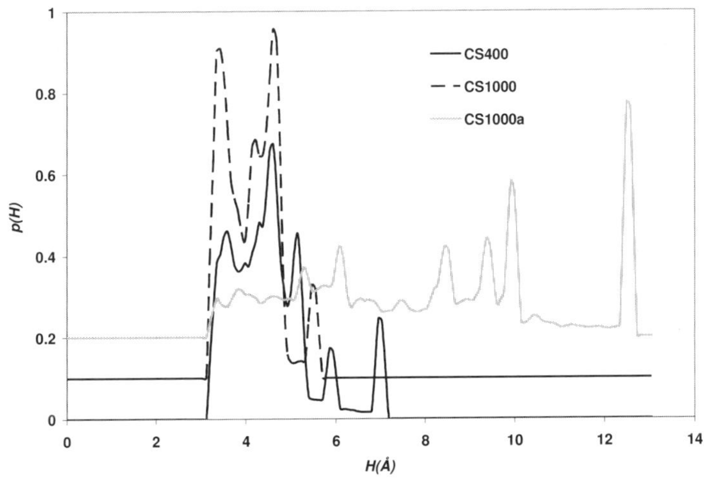

Figure 3. Pore size distribution of the three carbon models.

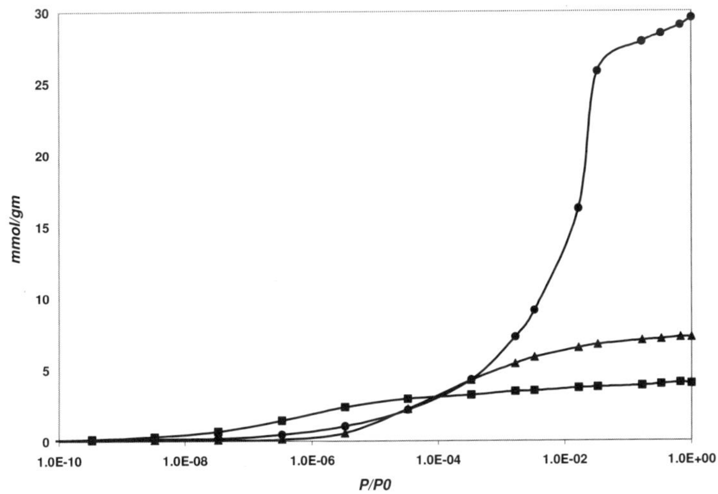

Figure 4. Argon adsorption isotherm at 77.4 K for models obtained using GCMC simulations in CS400 (triangles), CS1000 (squares) and CS1000a (circles).

We also calculated the argon adsorption at 77.4 K in the resultant models using GCMC simulations. All three adsorption isotherms shown in figure 4 (the x-axis has been plotted in log scale for clarity) are typical of microporous solids. We found that the amount adsorbed is much greater for CS1000a than for CS400 and CS1000. This is due to the high porosity of CS1000a as compared to the other two samples. Moreover, micropore filling starts at a lower pressure for CS1000 and CS400 as compared to CS1000a. This is due to the presence of narrow micropores in CS1000 and CS400. The micropore filling starts at a lower pressure for CS1000 as compared to CS400. This is due to the comparatively high density of carbon atoms in CS1000 as compared to CS400. Thus an adsorbate molecule in CS1000 feels the presence of a large number of carbon atoms as compared to the adsorbate in CS400.

4. Discussion

We have developed molecular models for 3 saccharose based carbons using a RMC method that incorporates an energy penalty term. The resultant models, as seen from the snapshots, reveal the disordered nature of porous carbons and have complicated pore geometry. The resultant molecular models reproduce the experimental pair correlation functions with good accuracy. The presence of the energy term in the acceptance criteria penalizes the formation of unphysical features such as 3 and 4 member rings and reproduces the correct local environment of the carbon atoms. Using a ring clustering method we found that the molecular models contain some defective graphene segments. Apart from the graphene segments, there are many carbon atoms which do not belong to any graphene segments and are arranged in a chain like fashion. The PSD reveals that our carbon samples consist mainly of micropores. CS400 and CS1000 have a narrow PSD, whereas CS1000a has a broad distribution. We also studied the adsorption of Argon in our molecular models. The adsorption isotherms are found to be typical of microporous solids for all the three models and we were able to rationalize the adsorption results on the basis of both PSD analysis and porosity.

Acknowledgements

SKJ thanks the French Ministry of Foreign Affairs for the award of an Eiffel Doctoral fellowship, and CNRS, Campus de Luminy, Marseille for their hospitality during the period when this work was carried out. We thank the Department of Energy (grant no. DE-FGO2-98ER14847) for support of this research. We thank the National Resource Allocation Committee of the National Science Foundation for a grant of supercomputer time.

References

1. McGreevy R. L. and Pusztai L., Reverse Monte Carlo simulation: a new technique for the determination of disordered structures, *Mol Sim* **1** (1988) 359-367.
2. Jain S. K., Fuhr J., Pellenq R. J-M., Pikunic J., Bichara C. and Gubbins K. E., Stability of porous carbon structures obtained from Reverse Monte Carlo using tight binding and bond order Hamiltonians, *Stud Surf Sci Catal* (in press).
3. Pikunic J., Clinard C., Cohaut N., Gubbins K. E., Guet J. M., Pellenq R. J-M., Rannou I. and Rouzaud J-N., Structural modeling of porous carbons: constrained Reverse Monte Carlo method, *Langmuir* **19(20)** (2003) 8565-8582.
4. Jain S. K., Gubbins K. E., Pellenq R. J-M. and Pikunic J., Molecular modeling of porous carbons using Hybrid Reverse Monte Carlo, *Langmuir* (submitted).
5. Opletal G., Petersen T., O'Malley B., Snook I., McCulloch D. G., Marks N. A. and Yarovsky I., Hybrid approach for generating realistic amorphous carbon structures using Metropolis and Reverse Monte Carlo, *Mol Sim* **28(10-11)** (2002) 927-938.
6. Petersen T., Yarovsky I., Snook I., McCulloch D. G. and Opletal G., Microstructure of an industrial char by diffraction techniques and Reverse Monte Carlo modeling, *Carbon* **42** (2004) 2457-2469.
7. Jain S. K., Pikunic J., Pellenq R. J-M. and Gubbins K. E., Effects of activation on the structure and adsorption properties of a nanoporous carbon using molecular simulation, *Adsorption* **11** (2005) 355-360.
8. Brenner D. W., Empirical potential for hydrocarbons for use in simulating the chemical vapor deposition of diamond films, *Phys Rev B* **42(15)** (1990) 9458-9471.
9. Tersoff J., Empirical interatomic potential for carbon, with applications to amorphous carbon, *Phys Rev Lett* **61** (1988) 2879-82.
10. Jain S. K. and Gubbins K. E., Ring Connectivity: Measuring network connectivity in network covalent solids, *Langmuir* (submitted)
11. Franzblau D. S., Computation of ring statistics for network models of solids, *Phys Rev B* **44(10)** (1991) 4925-4930.
12. Gelb L. D. and Gubbins K. E., Pore size distributions in porous glasses: a computer simulation study, *Langmuir* **15** (1999) 305-308.

CONTROLLING SELECTIVITY VIA MOLECULAR ASSEMBLING IN CONFINED SPACES: ALKANES – ALKENES - AROMATICS IN FAU ZEOLITES

J.F. DENAYER, I. DAEMS, G.V. BARON

Department of Chemical Engineering, Vrije Universiteit Brussel
Pleinlaan 2, B-1050 Brussel, Belgium
E-mail: Joeri.Denayer@vub.ac.be

PH. LEFLAIVE, A. METHIVIER

Institut Français du Pétrole - Lyon, BP n° 3, 69390 Vernaison, France

Liquid phase adsorption of alkane/alkene/aromatic mixtures in FAU supercages is governed by a combination of enthalpic and entropic effects. Large energetic interactions between specific molecular moieties (*e.g.* double bond or aromatic ring) and adsorption sites, lead to a preferential adsorption of aromatics compared to alkenes and alkanes. Entropic packing effects on the other hand are shown to be able to clearly outweigh normal tendencies for selectivity based on adsorbate properties (*e.g.* # C-atoms) and structural properties (*e.g.* aluminium contents) observed at low coverage. For the first time, it was shown that even in adsorbents or catalysts with relatively large pores, molecular selectivity is achieved at high degree of pore occupancy as a result of the assembly of molecules inside such pores. These selectivity effects, which are not acting at low degree of pore filling, depend in a subtle way on molecular size and shape, functional groups, pore size and geometry (*e.g.* spherical cage versus tubular pore), cation number and type, presence of solvents and so on. This concept of packing induced selectivity offers perspectives for new separation and catalytic processes.

1. Introduction

Selectivity is a key concept in catalytic and separation processes. It is a measure of the ability of a catalyst to convert one or more reagents into desired products, or for adsorptive separation processes, the ability of an adsorbent to remove a particular component from its mixture with other components. Selectivity is the key to better, more efficient and environmental friendly chemical processes. Even a small increase or reversal in chemical selectivity can transform a poorly performing process into an economically attractive one. The tools for controlling selectivity are: a careful tuning of active sites such as cation type and amount, promoters, chemical properties and structure of the support or material, solvents and operating conditions.

Microporous solids [1,2], with their nanosized pores, show very high catalytic activity and adsorption capacity as a result of their very large internal surface area. Such materials furthermore may possess a unique property called shape-selectivity, which is the ability to discriminate between molecules based on their molecular size or shape. Classical shape-selectivity is limited to systems with pores having dimensions very similar to those of the invited molecules:*e.g.* 10 membered ring zeolites such as ZSM-5, ZSM-22 and ZSM-23 or materials with narrow windows between the cages such as LTA, e.g. zeolite 5A [3-5]. Often, the selectivity results from some molecules being able to enter (linear hydrocarbon) and others not (branched hydrocarbon). More subtle effects and even inverse shape selectivity (preference for the branched molecule) can result from entropic or ordering effects in these materials [6-9]. In gas phase, there is a strong dependence of the amount adsorbed on the chain length or size of the molecule [10], a dependence which usually disappears in liquid phase or at high loading (where most industrial operations operate) and generally, selectivity is lost for large pore materials [11, 12]. Selectivity is however largely retained for small pore materials where interaction with the zeolite channel walls dominates over intermolecular interactions [2, 4, 13, 14].

For molecules which differ in size or shape and electrostatic interactions such as the xylene isomers [15], liquid phase separations can be performed and selectivities tuned on FAU zeolites by adequate choice of the compensating cations. Other cases still allowing separation are to be found in large pore materials presenting sub-cavities such as with MCM-22 or biporous materials [8].

In many hydrocarbon separations, molecules in the mixture are so very similar in size, shape and other properties that a simple change of interaction with a cation or pore size does not yield a useful selectivity. Very small differences have to be exploited to still obtain a separation and the driving force is then mainly based on differences in ordering the molecules in the mixture in a confined space, eventually enhanced or controlled by adding a solvent to the mixture of adequate size and shape.

Apart from selectivity, *capacity* is a crucial parameter for separating agents, as is *activity* for catalysts. Capacity and activity largely influence the size of equipment and cost of industrial separation and catalytic processes. Capacity and activity are proportional to the contact surface between the molecules and the catalyst adsorbent, which in turn is inversely proportional to the catalyst/adsorbent pore diameter. Disadvantages of solids having such small pores is that (i) diffusion is severely slowed down in their pore system and (ii) they cannot accommodate many of the larger molecules found in chemical feedstocks, limiting their field of application. Zeolites with larger pores circumvent these disadvantages, but unfortunately, such materials are almost invariably unselective according to scientific and patent literature.

In this paper, we review some of our recent work [16-20], performed to investigate whether selectivity can still be obtained in such solids with larger pore systems via molecular assembling mechanisms. *Molecular assembling* can be defined as the arrangement of adsorbed molecules inside confined pore systems, hereby optimizing the balance between energetic and steric contributions. Such packing effects are obviously only important at a high degree of pore filling [3, 13]. Remarkably, very few scientific publications [21-25] discuss adsorption in microporous solids in such conditions, where however most industrial processes operate. As an example, we will discuss the case of liquid phase alkane – alkene – aromatic separation in FAU zeolites such as NaX and NaY type zeolites.

2. Materials and methods

The performance of FAU zeolites critically depends on their Si:Al ratio, or cation content and cation type. X zeolites (Si:Al 1-1.5) have a higher aluminum contents than Y zeolites (Si:Al 1.6-3), but possess the same open 3-dimensional crystal structure. This structure [2, 26] consists of sodalite cages (β-cages) and hexagonal prisms that are connected in such a way that large internal supercages (α-cages) are created (Figure 1). Relatively large molecules can enter the α-cages through 12 Membered Ring (12MR)-windows without being sterically hindered. Therefore the classical shape selectivity does not occur on this material. Cations positioned on sites II (SII) and III/III' (SIII/III') are exposed inside the supercages and are considered to be the most important adsorption sites for polar molecules. SII and SIII are located respectively near the 6-ring of the β-cage and the 4-ring of the β-cage. SIII' is closely related to SIII, but positioned inside the 12MR-window.

The NaX and NaY zeolite samples used for the liquid phase experiments were provided by Institut Français du Pétrole (IFP) and had the typical Si:Al-ratio of 1.23 and 2.79 respectively, as given in Table 1. The Dubinin micropore volumes were determined by means of N_2-porosimetry. The theoretically available micropore volume per g zeolite for hydrocarbon adsorption, 0.32 ml/g, was calculated by multiplying the total volume of the supercages per unit cell (UC) (6700Å^3) [27], with the total number of unit cells per g zeolite. This available volume for the hydrocarbons is lower than the Dubinin micropore volume since N_2 molecules can enter both α- and β-cages, while hydrocarbons can exclusively enter α-cages. The maximum available volume for hydrocarbon adsorption (*e.g.* benzene) is about 0.3 ml/g for both NaX and NaY as there are complex interactions with the space occupied by cations, their attraction and ordering of the molecules in the remaining space. When replacing Na by say Cs, a much larger cation, even less space is available.

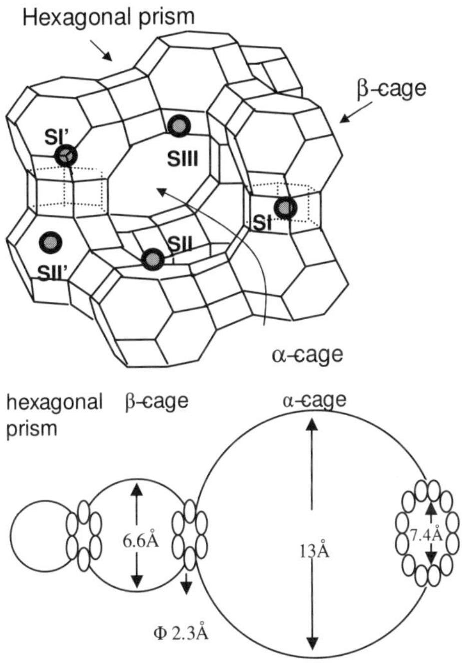

Figure 1. Structure of faujasites X and Y with cation positions SII and SIII in the supercages, SI' and SII' in the β-cages and SI in the centers of the hexagonal prisms. Dimensions of faujasite windows and cages.

Table 1. NaX and NaY zeolite material properties and Henry law coefficients for n-hexane and benzene

K' (mol/(kg Pa))	Si:Al	N_2 Micropore volume (ml/g)	n-C6	Benzene
NaX	1.23	0.31	8.54E-05	9.46E-04
NaY	2.79	0.35	3.89E-05	1.69E-04

Experimental details of the batch method used to determine binary adsorption isotherms were previously described [16]. In the batch technique, a known amount of mixture of the component(s), eventually in a solvent are contacted with adsorbent and from an analysis of the external phase after equilibration and a mass balance, the amount adsorbed is calculated. In a two component mixture, one is limited to low concentrations of the adsorbates, as the amount of adsorbate added to the zeolite can not largely exceed the available micropore volume, in order to be able to accurately detect changes in the concentration upon adsorption. Data are at room temperature (20°C) unless otherwise noted.

3. Liquid phase adsorption of alkene-alkane mixtures

The adsorption of alkenes with different chain length (C_6-C_{12}) from alkane solvents (C_5-C_{14}) on NaY (Si:Al 2.79) was studied using a batch experimental technique. Under these conditions the zeolite micropores are close to saturation, since the solvent (alkane) will show a tendency to fill up the remaining free space. Already at low alkene concentrations, the alkenes are selectively adsorbed from their mixture with an alkane as a result of the specific interactions between π-electrons of the double bond and zeolite cations. The amount alkene adsorbed depends on the chain length of both the alkene and the alkane solvent in an unexpected way. Two remarkable effects are observed: (1) shorter alkenes are preferentially adsorbed compared to longer alkenes and (2) with longer alkane solvents, the hexene/dodecene selectivity decreases (Figure 2).

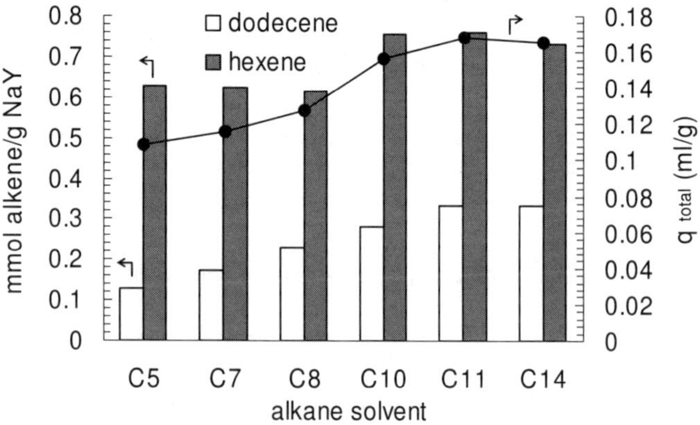

Figure 2. Liquid phase adsorption of an equimolar hexene/dodecene mixture (2 mol% each) from different alkane solvents (96 mol%) on NaY (Si:Al 2.79).

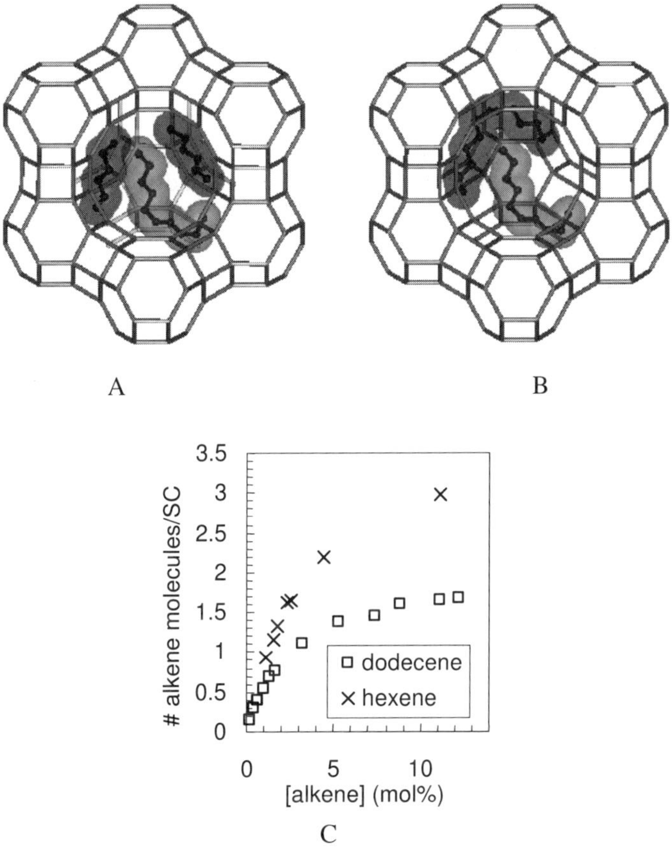

Figure 3. Schematic presentation of the co-adsorption of (A) hexene and decane and (B) dodecene and decane in a NaY supercage at an external alkene concentration of 3 mol%. (C) Adsorption isotherms of hexene and dodecene from their mixture with decane on NaY (Si:Al 2.79).

These observations are completely different from the usual increase in adsorption strength or selectivity with increasing carbon number as observed in diluted gas-phase conditions [10]. Apparently, shorter linear hydrocarbons, having a smaller number of C-atoms pack more efficiently at higher degree of pore filling and are in other words favorably adsorbed because they can easily fill gaps within the zeolite matrix, as illustrated in Figure 3A-B. In the adsorption of hexene and dodecene from their mixture with decane, the empty space next to the adsorbed decane solvent molecule can be filled with either 2 hexene or 1 dodecene molecule(s). Entropically, the adsorption of 2 hexene molecules is more favorable than the adsorption of only 1 dodecene molecule, leading to the preferential adsorption of hexene (Figure 3C). This effect was not really expected to occur on large cage-type zeolites capable of hosting multiple molecules per supercage.

The more efficient packing of small alkenes is found to become even more pronounced with increasing alkene loading, as shown in Figure 4 with batch adsorption data of equimolar mixtures of hexene and dodecene dissolved in heptane on NaY. While the amount dodecene adsorbed remains more or less constant with increasing alkene concentration, the amount hexene adsorbed drastically increases.

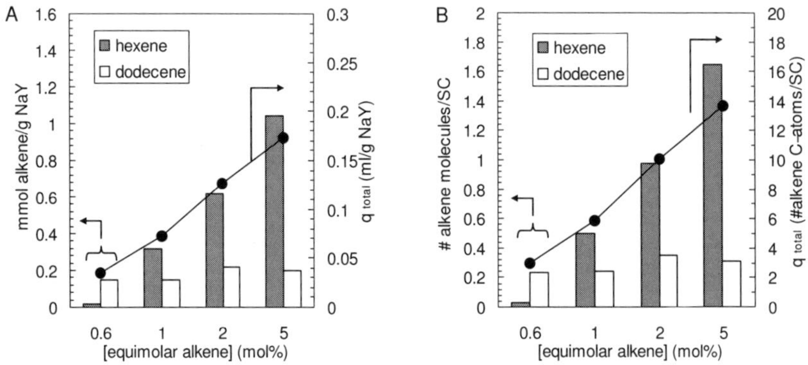

Figure 4. Adsorption of equimolar hexene/dodecene mixtures on NaY (Si:Al 2.79) as a function of alkene concentration in the solvent heptane.

In these very low bulk alkene concentration conditions where these experiments are possible (Figures 2 and 4), it should be noted that the micropores are at maximum (C_{11}-C_{14} or high concentration) already filled with alkenes for about 50 - 60%, and clearly this increases with alkene concentration. There is hence a high selectivity (up to 6.1 in heptane at 5 mol%) towards the shorter alkenes in mixtures as shown in Table 2.

Table 2. Selectivity factors (α_{hd}) for equimolar hexene/dodecene mixtures adsorbed from heptane and undecane on NaY (Si:Al 2.79).

Solvent	0.6 mol%	1 mol%	2 mol%	5 mol%
heptane	-	2.2	4.0	6.1
undecane	2.7	3.2	2.7	5.6

In absence of alkane solvent (Table 3), this highly non-ideal behavior leads even to a separation factor higher than 9 for a hexene/dodecene mixture allowing their very efficient separation. In practice, the above mentioned packing effects for alkane/alkene mixtures can be exploited in adsorptive separation or catalytic processes: the relative selectivity for alkenes with different chain length can be adjusted by choosing different alkane solvents and different alkene concentrations.

4. Liquid phase adsorption of aromatics

Normally cation type and amount are used to tune the selectivity for aromatic compounds (*e.g.* xylenes). Additionally, unexpected packing induced selectivity effects were observed for the liquid phase adsorption of aromatics. The adsorption of benzene, toluene, m-xylene and mesitylene from their binary mixtures with octene or octane was studied on Na-FAU having different Si:Al-ratios. It was found that NaY (Si:Al 2.79; low cation content) is a more selective adsorbent compared to NaX (Si:Al 1.23; high cation content). As an example, the data for benzene are given in Figure 5. Furthermore, no differences were observed between the adsorption of aromatics on NaX and LSNaX (Si:Al 1.02; very high cation content).

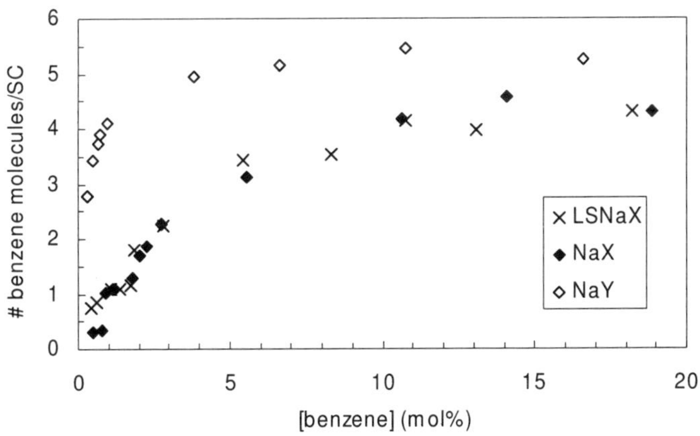

Figure 5. Quantity of benzene adsorbed from octene on zeolites LSNaX (Si:Al 1.02), NaX (Si:Al 1.23) and NaY (Si:Al 2.79) in liquid phase at room temperature.

The observation that a high-silica zeolite is found to adsorb the aromatic compound more selectively compared to its low-silica counterpart is in clear contrast to what is typically observed for pure aromatics in gas phase. Table 1 also gives the Henry law coefficients for n-hexane and benzene on NaX and NaY, and clearly, increasing the cation content increases the amount adsorbed strongly for n-hexane and dramatically for aromatics such as benzene in gas phase.In gas phase conditions however, the zeolite pores only contain aromatics, often at low degree of pore occupancy, while under the present conditions the pores are close to saturation and contain benzene as well as solvent molecules. Cations on SIII/III' (absent in NaY supercages) are believed to cause a skewed docking of aromatics on NaX SII sites because of their electrostatic interactions with the π-electrons of aromatic ring structures. Such an orienting effect leads to a large entropic disadvantage in crowded supercages. This hypothesis is completely in line with the fact LSNaX shows the same selectivity for all studied aromatics as NaX. Supercages of NaX already contain 4 SIII/SIII' cations that influence the adsorption of aromatic molecules on each SII site (4 per NaX supercage). Therefore the presence of additional SIII/SIII' cations will not lead to a further decrease of the selectivity on LSNaX. Furthermore SIII/SIII' cations probably hamper the van der Waals interactions between benzene and the zeolite framework, thereby disturbing the accommodation of benzene inside the 12MR-window of NaX (Figure 6).

This "reverse" behaviour with respect to cation content at high pore occupancy is not a rule (Figure 7). For the adsorption of alkenes, the influence of the Si:Al-ratio is in line with what could be expected from observations at low coverage. Alkenes are found to be more selectively adsorbed on NaX than on NaY. As for the practical consequence of these observed selectivity patterns, despite it's lower cation content, NaY is proven to be a better separation agent for alkane/alkene/aromatic mixtures compared to NaX.

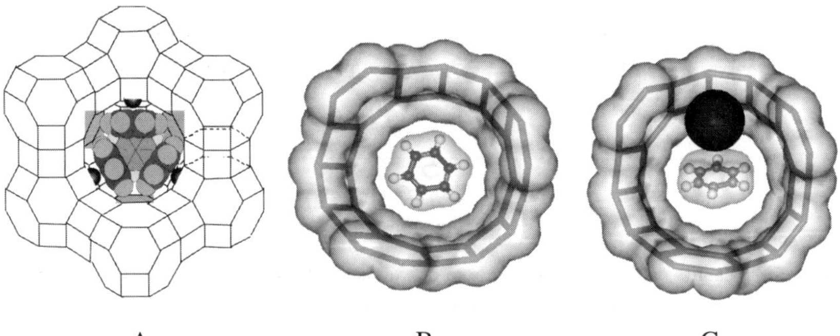

A B C

Figure 6. Schematic representation of the adsorption of 4 benzene molecules on the SII sites, and 5th one in the 12MR-window of (B) NaY in the absence of SIII/SIII' cation and (C) NaX in the presence of SIII/SIII' cation.

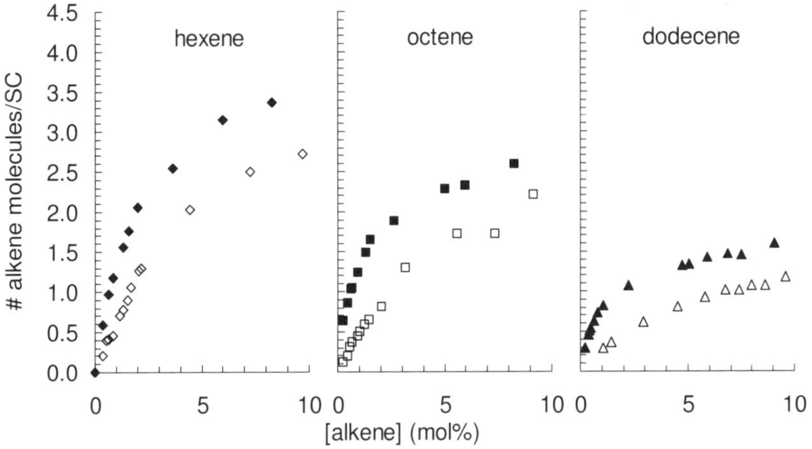

Figure 7. Quantity of hexene, octene and dodecene adsorbed from heptane on zeolites NaX (Si:Al 1.23; full symbols) and NaY (Si:Al 2.79; empty symbols) in liquid phase at room temperature.

5. Practical applicability of molecular assembly effects

Petroleum fractions contain many different hydrocarbon molecules and ever more stringent environmental constraints now determine composition and purity requirements of the products. Furthermore, when upgrading different hydrocarbon streams the formation of side-products leads to even more complex mixtures. For example when producing linear olefinic hydrocarbons by paraffin dehydrogenation aromatic side-products are formed [28]. Often, alkane/alkene/aromatic hydrocarbon mixtures have to be separated. For the liquid phase separation of *normal* alkenes from *n*-alkene/*n*-alkane mixtures, the OLEX process was developed [2]. Also, the separation of alkane/alkene mixtures by adsorption via π-complexation has been extensively studied [29-31]. However, no industrial adsorptive separation processes are available for the separation of either alkanes or alkenes of different chain length. Rather, a downstream distillation section is used as to separate for example the linear *alpha*-olefins (C_4-C_{10}) produced by the AlphaSelect Process (IFP) [32].

Given the large number of hydrocarbon mixtures in the petrochemical industry that have to be separated, there is still a large growing potential for adsorptive separations. Two examples are given next to illustrate the applicability of FAU type zeolites for the separation of (i) alkenes with different chain length and (ii) alkane/alkene/aromatic mixtures with data from actual column separation tests.

5.1. *Separation of short and long chain alkenes*

A column separation experiment with heptane solvent carrier, containing an equimolar hexene/dodecene (both 2 mol%) mixture was performed. The column had an internal volume of 0.77 cm^3 and contained 0.443 g NaY (Si:Al 2.79). Figure 8 shows the break-through profiles of this hexene / dodecene / heptane mixture at room temperature. Heptane is weakly retained by the adsorbent and elutes directly (not shown in graph). Clearly, hexene is retained longer in the column compared to dodecene, which is in accordance to the results obtained in the batch experiments. Breakthrough of dodecene is observed after just 4.5 minutes, whereas hexene only starts to elute after 9 minutes.

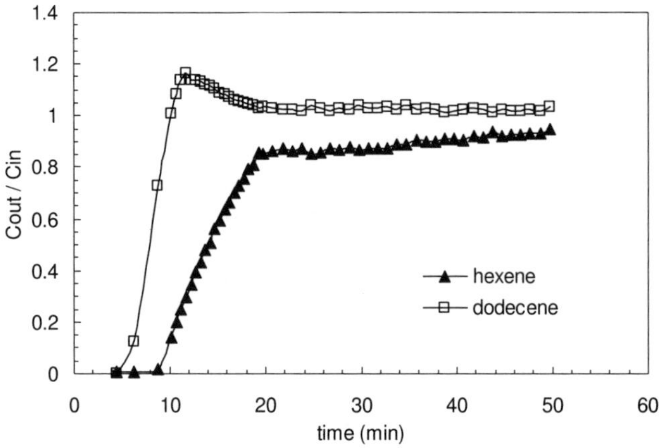

Figure 8. Breakthrough profiles of an equimolar hexene/dodecene mixture (both 2 mol%) using heptane (96 mol%) as solvent at room temperature on a column (0.77 cm3) packed with NaY crystals (Si:Al 2.79; 0.443g).

Our batch results showed an unexpected increase in selectivity towards the shortest alkene with increasing external alkene concentration and this is also observed in column separations. This non ideal behavior was further investigated by performing experiments on a pilot scale breakthrough set-up using a column with an internal volume of 86.4 cm^3 containing binderless NaX beads (Si:Al 1.33). These experiments were performed at IFP (Lyon). Heptane (solvent) containing a hexene/dodecene mixture having equal weight fractions was pumped through the column. Figure 9 shows the breakthrough curve of a 10% hexene / 10% dodecene / 80% heptane mixture at 50°C. Dodecene leaves the column before hexene and thus is less adsorbed than hexene. Similar experiments were performed on the same set-up with other ternary hexane /

dodecene / heptane mixtures containing respectively 2, 4, 30 and 50 weight percentage of both alkenes. Calculation of the mass balance allows the determination of the amounts of hexene and dodecene adsorbed inside the micropores of NaX. Figures 10 A-B show the evolution of the amounts hexene and dodecene adsorbed in function of their concentration (weight %) in the liquid feed. The total alkene volume adsorbed increases with the alkene concentration. In absence of heptane solvent, the alkenes occupy the total internal volume of NaX (0.35 ml/g or 25 alkene C-atoms/SC).

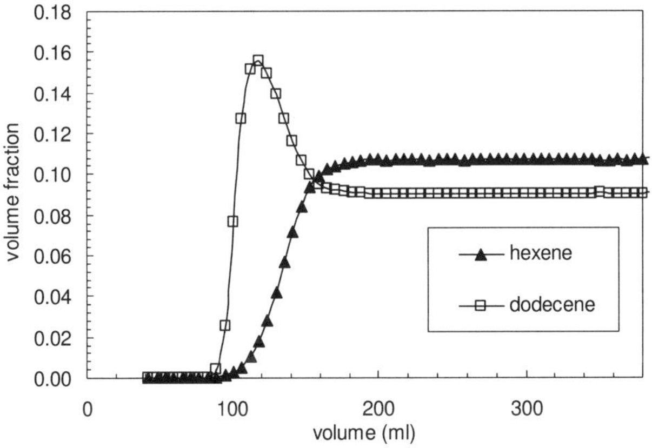

Figure 9. Breakthrough profiles of ternary hexene/dodecene/heptane (10/10/80 weight %) mixture at 50°C on a column (86.4 cm3) packed with binderless NaX beads (56.24 g).

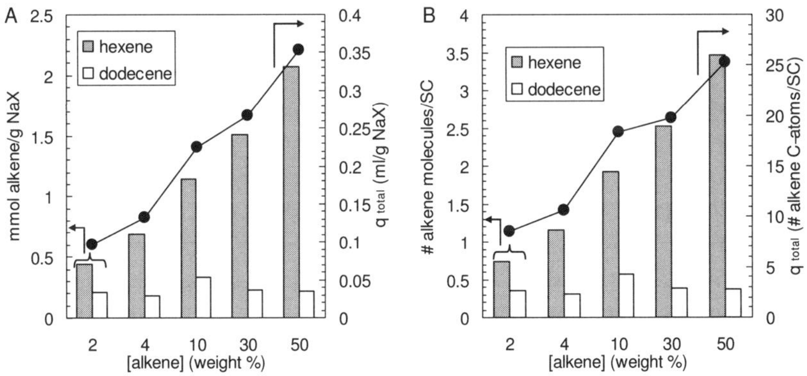

Figure 10. Amounts hexene and dodecene adsorbed from heptane solvent in pilot scale breakthrough experiments on NaX at 50°C with different alkene feed concentrations.

While the amount hexene adsorbed increases with its concentration, the amount of dodecene is not affected by its concentration in the bulk phase, in agreement with the batch adsorption experiments presented above (Figure 4). The same trends were observed when using a different solvent such as decane. Selectivity factors of hexene over dodecene adsorbed from heptane are given in Table 3.

In agreement to what was observed in the ternary batch adsorption experiments (Table 2), the separation factor increases with increasing alkene loading. In the co-adsorption of the 50-50% hexene/dodecene solvent free mixture, a separation factor as high as 9.2 was obtained. Such a separation factor is large enough to allow bulk phase separation of these components.

Table 3. Selectivity (α_{hd}) of hexene/dodecene from pilot scale breakthrough experiments at 50°C on NaX

Alkene wt %	2	4	10	30	50
α_{hd}	2.2	3.8	3.2	6.9	9.2

5.2. *Column separation of octene and benzene: influence of Si:Al*

The Si:Al-ratio of Na-FAU has an opposite effect on the adsorption selectivity of aromatics and alkenes in liquid phase: while NaX has a higher selectivity for alkenes compared to NaY (Figure 7), NaY has a higher selectivity for aromatics than NaX (Figure 5). This selectivity pattern is schematically represented in Figure 11.

Breakthrough experiments were performed in order to verify this hypothesis. Heptane (96 mol%), containing an equimolar octene/benzene (both 2 mol%) mixture, was separated on columns (with identical dimensions) containing NaX (0.536 g) and NaY (0.4295 g) respectively. The breakthrough profiles are presented as a function of the liquid feed volume per g adsorbent that was pumped through the column (Figure 12). With NaX, octene elutes after 6.5 ml feed/g zeolite. This is later compared to NaY, where the alkene elutes the column after 4.5 ml feed per g NaY. On the other hand, benzene leaves the NaY column after 22 ml/g compared to 15.8 ml/g NaX. The volume of liquid mixture per g zeolite that passes the column after the breakthrough of octene and before the breakthrough of benzene, 17.5 ml/g NaY compared to 9.3 ml/g NaX, is clearly larger for NaY compared to NaX, making NaY a better adsorbent to separate alkenes from aromatics compared to NaX.

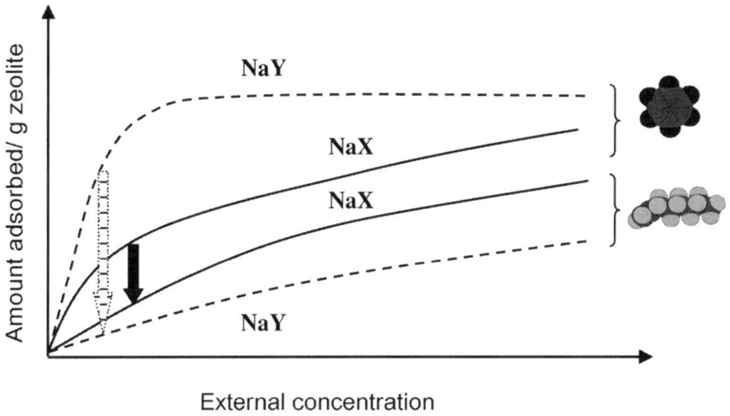

Figure 11. Schematic representation of the binary adsorption isotherms of benzene and octane adsorbed from heptane on NaX (black lines) and NaY (dotted lines).

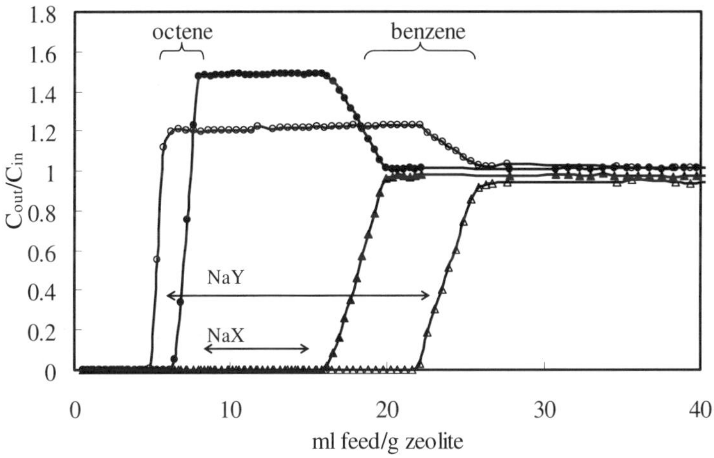

Figure 12. Breakthrough profiles of equimolar benzene/octene (2 mol%) mixture using heptane solvent at room temperature on columns (0.85 cm3) packed with NaX (full symbols) and NaY (open symbols) crystals.

6. Conclusions

Nowadays we dispose of a large number of zeolites which can separate mixtures of alkanes, alkenes and aromatics based on shape selectivity or specific interactions with cations. Unfortunately, many of these materials have very small pore volumes and hence capacities, limiting or preventing their economical feasibility in large scale bulk liquid phase separation processes. In this work we have demonstrated that there is however a large potential for exploiting molecular assembling effects (entropic rather than enthalpic or energy effects) in traditional low cost large pore zeolite materials.

Acknowledgements

This research was financially supported by Institut Français du Pétrole. J. Denayer is grateful to the F.W.O.-Vlaanderen, for a fellowship as postdoctoral researcher.

References

1. Guisnet M., Gilson J.-P. (Eds.), *Zeolites for Cleaner Technologies* (Catalytic Science Series, 3), ISBN: 1860943292, (World Scientific Publishing Company, 2002).
2. Sherman J.D., Proc. Natl. Acad. Sci. USA, 96 (1999) pp. 3471.
3. Denayer J.F.M., De Meyer K.M.A., Martens J.A., Baron G.V., Angew. Chem. Int. Ed., 42 (2003) pp. 2774-2777.
4. Ocakoglu R. A., Denayer J.F.M., HuybrechtsW., Marin G. B., Martens J.A., Baron G.V., J. Phys. Chem. B, 207 (2003) pp. 398-406.
5. Denayer J. F., Ocakoglu A. R., Huybrechts W., Martens J. A., Thybaut J. W., Marin G. B., Baron G. V., Chem. Comm. (2003) pp. 1880-1881.
6. Eder F., Lercher J.A., Zeolites, 18 (1997) pp. 75.
7. Santilli D.S., Harris T.V., Zones S.I., Microporous Materials, 1 (1993) pp. 329-341.
8. Denayer J.F.M., Ocakoglu R.A., Arik I.C., Kirschhock C.E.A., Martens J.A., Baron G.V., Angew. Chem. Int. Ed., 44 (2005) pp. 400-403.
9. Denayer J.F.M., Ocakoglu R.A., De Meyer, K., Baron, G.V., Adsorption, 11 (2005) pp. 49-53.
10. Denayer J.F.M., Baron G.V., Adsorption, 3 (1997) pp. 251.
11. Denayer J.F.M., Bouyermaouen A., Baron G.V., I&EC, 37 (1998) pp. 3691-3698.
12. Denayer, J.F.M., De Jonckheere, B., Hloch, M., Marin, G. B., Vanbutsele, G., Martens, J.A., Baron, G.V., J. Catal., 210, (2002) pp. 445-452.
13. Chempath, S., de Meyer, K., Denayer, J.F.M, Baron, G.V., Snurr, R.Q., Langmuir, 20 (2004) pp. 150-156.
14. Denayer J.F.M., Ocakoglu A.R., Martens J.A., Baron G.V., J. Catal., 226 (2004) pp. 240-244.
15. Kulprathipanja S., Johnson J.A. in: F. Schüth, K.S.W Sing, J. Weitkamp (Eds.), Handbook of porous solids, (Wiley-VCH, Weinheim, 2002) pp. 2568-2612.

16. Daems I., Leflaive Ph., Méthivier A., Denayer J.F.M. and Baron G.V., Adsorption, **11** (2005) pp. 189-194.
17. Daems I., Leflaive Ph., Méthivier A., Denayer J.F.M. and Baron G.V., Microporous and Mesoporous Materials, **82** (2005) pp. 191-199.
18. Daems I., Méthivier A., Leflaive Ph., Fuchs A.H., Baron G.V. and Denayer J.F.M., *Journal of the American Chemical Society*, **127** (2005) pp. 11600-11601.
19. Daems I., Leflaive Ph., Méthivier A., Baron G.V., Denayer J.F., Influence of Si:Al ratio of Faujasites on the Adsorption of Alkanes, Alkenes and Aromatics, submitted Microporous and Mesoporous Materials (2006)
20. Denayer J. F.M. , Daems I., Baron G.V., Adsorption and Reaction in Confined Spaces, Proceedings of the "Research Advances in Rational Design of Catalysts and Sorbents" conference, to appear in *Oil & Gas* Science and Technology - Revue de l'IFP (2006)
21. Iwayama K. and Suzuki M., *Studies in Surface Science and Catalysis*, **83** (1994) pp. 243.
22. Krishna R. , *Chemical Engineering Research & Design*, **79** (2001) pp. 182.
23. Denayer J.F.M., Ocakoglu R.A., Huybrechts W., Dejonckheere B., Jacobs P., Calero S., Krishna R., Smit B., Baron G.V., Martens J.A., *Journal of Catalysis*, **220** (2003) pp. 66.
24. Chiang A.S.T., Lee C.K., Chang Z.H., *Zeolites*, **11** (1991) pp. 380.
25. Krishna R., Smit B., Vlugt T.J.H., *J. Phys. Chem. A*, **102** (1998) pp. 7727.
26. Baerlocher Ch., Meier W.M., Olson D.H., *Atlas of zeolite framework types*, (Elsevier, Amsterdam, 2001).
27. Breck D.W., Zeolite Molecular Sieves: structure, chemistry and use, (John Wiley & Sons, New York, 1974).
28. Vora B.V., U.S. Patent 5,300,715, assigned to UOP (1994).
29. Yang R.T. and Kikkinides E.S., AIChE Journal, **41**, (1995) pp. 509.
30. Rege S.U., Padin J., Yang R.T., AIChE Journal, **44**, (1998) pp. 799.
31. Padin J., Yang R.T., Munson C.L., Ind. Eng. Chem. Res., **38**, (1999) pp. 3614.
32. Bourbigou H.O., Chodorge J.A., Travers Ph., Pet. Technol. Quart., **Q4**, (1999) pp. 141.

A NEW METHODOLOGY IN THE USE OF SUPER-CRITICAL ADSORPTION DATA TO DETERMINE THE MICROPORE SIZE DISTRIBUTION

D. D. DO, H. D. DO AND G. BIRKETT

Chemical Engineering, University of Queensland, St. Lucia, Qld 4072, Australia
E-mail: duongd@cheque.uq.edu.au

Adsorption of methane on the surface of graphitized thermal carbon black and in slit pores is studied using the method of Grand Canonical Monte Carlo simulation. Under the supercritical conditions and very high pressure the mass excess decreases towards zero value for a graphite surface, while for slit pores negative excess density is possible at extremely high pressures. Adsorption data obtained under supercritical conditions are increasingly used to determine the pore size distribution in the micropore range. This is largely motivated by the advances in the use of supercritical adsorption in high energy applications, such as hydrogen and methane storage in porous media. Experimental data reported as mass excess versus pressure and when these data are matched against the theoretical mass excess, significant errors can occur if the void volume used in the calculation of the mass excess is incorrectly determined. The incorrect value for the void volume leads to wrong description of the maximum in the plot of mass excess versus pressure and the part of the isotherm over the pressure region where the mass excess decreases with pressure. Because of this uncertainty in the maximum, we propose a new method in which the problems associated with this maximum of the surface excess are completely avoided. Our method involves only the relationship between the amount that is introduced into the adsorption cell and the equilibrium pressure. This information of "direct" experimental data has two distinct advantages. The first is that the data is the direct data without any manipulation (i.e. involving further calculations), and the second one is that this relationship is always monotonically increasing with pressure. We will illustrate this new method with the adsorption data of methane in a commercial Ajax activated carbon.

1. Introduction

Adsorption of gases on non-porous surfaces and in porous solids has been increasingly studied by Monte Carlo, Molecular Dynamics and Density Functional Theory methods [1-5]. Equilibria of simple gases is now routinely studied with these methods, and the predicted adsorption isotherms generally agree well with experimental data of well-defined surfaces, such as graphitized thermal carbon black (GTCB) [6-8]. However, the success of the predictions depends on the choice of the potential equations and the correct selection of the molecular parameters. For the case of methane, it is often modeled as a pseudo-spherical particle with one interaction center although in the confined space of pores one would expect that the orientation of methane molecule should play an important role in adsorption. Therefore, it is expected that the 5-Site model is more appropriate than the equivalent 1-Site model in the description of adsorption in pores because it is known to be a better model to describe liquid and solid behaviors [9-10]. Since methane is one of the high-energy gases and its potential utilization by storage in porous materials at high pressure is used as an alternative to gasoline, it is important to determine the pore size distribution (PSD) with methane and the question is raised about whether the 1-Site potential model for methane is as good as the 5-Site model in the determination of PSD. A new method for determining PSD is also developed to avoid the common problems associated with the reported excess density versus pressure.

2. Theory

Although there are many models that have been proposed for methane in the literature, we choose the 1-Site model suggested by Martin and Siepmann [11] and the 5-Site model of Chen and Siepmann [12] because these models describe well the vapor-liquid equilibria. In our previous publication [13] in which we evaluate the performance of the 1-Site and 5-Site model of Sun et al. [14] on the description of adsorption of methane on graphite and in graphitic pore, we have found that these models describe well the adsorption on open surfaces but they differ in the description of adsorption in graphitic slit pores, emphasizing the importance of the 5-Site model to account for correctly the molecular shape in the confined space. The 5-Site model of Sun et al. has five dispersive sites and five electrostatic charges. Since the effect of charge is insignificant, we shall use in this paper the recent 5-Site model of Chen and Siepmann (CS), which contains only five dispersive charges, to determine PSD.

The potential energy of site-site interaction follows the LJ equation:

$$\varphi_{i,j}^{(a,b)} = 4\varepsilon^{(a,b)}\left[\left(\frac{\sigma^{(a,b)}}{r_{i,j}^{(a,b)}}\right)^{12} - \left(\frac{\sigma^{(a,b)}}{r_{i,j}^{(a,b)}}\right)^{6}\right]$$

which describes the potential energy between the site *a* on the particle *i* and the site *b* on the particle *j*. Knowing the site-site interaction energy, the potential energy of interaction between two particles is simply obtained by summing all the pairwise potentials and assuming pairwise additivity. The molecular parameters are listed in Table 1.

Table 1. Molecular parameters for the 1-Site and 5-Site models

	σ (A)	ε/k_B (K)	Reference
1-Site Model			
CH_4	3.73	148	Martin and Siepmann [11]
5-Site Model: The C-H bond length of 1.1 A, and the angle H-C-H is 109.5 degrees.			
C	3.31	0.01	Chen and Siepmann [12]
C-H site	3.31	15.3	

The solid-fluid potential between a site of methane molecule and the surface is assumed to take the form of Steele 10-4-3 equation [15]. For the five-site model, each of the five sites interacts with the graphite surface in the same way. The solid-fluid well-depth of interaction energy is calculated with the following equation, $\varepsilon^{(a,b)} = (1-k)\sqrt{\varepsilon^{(a,a)} \times \varepsilon^{(b,b)}}$, where *a* and *b* to denote methane and carbon, respectively, and *k* is the binary interaction parameter and we assume that this binary interaction parameter is the same for all five interaction sites. The well-depth for carbon atom in the graphite is 28 K. The Steele 10-4-3 equation describes the interaction between an interaction site *a* of a fluid particle *i* and a graphitic lattice with its sub-lattices, and it is given by:

$$\varphi^{(a,b)}_{i,\text{lattice with sub-lattices}} = \varphi_w \left[\frac{1}{5}\left(\frac{\sigma^{(a,b)}}{z_i^a}\right)^{10} - \frac{1}{2}\left(\frac{\sigma^{(a,b)}}{z_i^a}\right)^{4} - \frac{\left(\sigma^{(a,b)}\right)^4}{6\Delta(z_i^a + 0.61\,\Delta)^3} \right]$$

Here the wall potential parameter φ_w is given by $\varphi_w = 4\pi\rho_s\varepsilon^{(a,b)}\left[\sigma^{(a,b)}\right]^2$, where ρ_s is the density of carbon atom per unit surface area of the graphite layer (ρ_s = 0.382 A^{-2}). The collision diameter of carbon atom in graphite layer is 3.4 A.

2.1. *Grand Canonical Monte Carlo simulation*

The molecular simulation method employed in this paper is the Grand Canonical Monte Carlo (GCMC) simulation. The parameters associated with the MC simulation used in this paper are (i) the linear dimension of the simulation box in the x- and y-directions is at least 10 times the collision diameter, (ii) the cut-off radius is taken to be half of the box length, (iii) the number of cycles for equilibration and statistical collection step is 50,000 and (iv) in each cycle, there are N displacement moves and N rotations (in the case of 5-Site model) where N is the number of particles in the simulation box. The simulation box is constrained in the x and y directions by the periodic boundary conditions.

From the GCMC simulation, we can obtain the isosteric heat as follows [1]:

$$-\Delta h = R_g T - \frac{\langle U_{a,ext} N \rangle - \langle U_{a,ext} \rangle \langle N \rangle}{\langle N^2 \rangle - \langle N \rangle^2}$$

where $U_{a,ext}$ is the potential energy between adsorbate molecules plus that between adsorbate molecule and the solid substrate. The potential energy of interaction can be broken down into contributions of fluid-fluid interaction and fluid-solid interaction.

3. Results and Discussion

3.1. *Adsorption on Graphitized Thermal Carbon Black under Sub-Critical Conditions*

To establish the adequacy of the CS-5-Site model, we use the experimental data of methane for adsorption capacity on graphitized thermal carbon black of Avgul and Kiselev [6]. The carbon black used by these authors is a highly homogeneous graphitized thermal carbon black, Sterling FT (2800), which had been obtained from Cabot Corporation. The N_2-BET surface area of this sample is 12.22 m^2/g. The adsorption data are fairly extensive and suitable to test the capability of the model for their prediction of adsorption isotherms. The results of the GCMC simulations are shown in Figure 1, where we plot the surface excess (mol/m^2) versus pressure. The experimental data are presented as symbols, while the results from the 1-Site model are shown as the dashed line and those from the 5-Site CS model as the solid-line. The binary interaction parameters for the 1-Site model and the 5-Site model are reasonably independent of temperature and these are listed in Table 2.

Table 2. The binary interaction parameters at various temperatures for the 5-Site and 1-Site models

T (K)	k (5-Site model)	k (1-Site model)
113	-0.05	-0.03
123	-0.05	-0.032
133	-0.06	-0.032
143	-0.06	-0.04

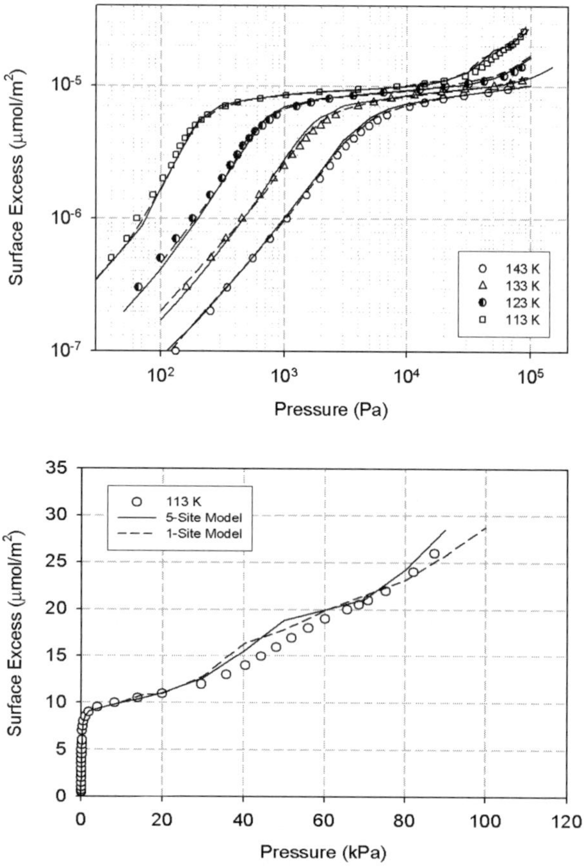

Figure 1. (LEFT) Adsorption isotherm of methane on graphitized thermal carbon black at 113, 123, 133 and 143 K (Experimental data: symbols; 5-Site model: solid line; 1-Site model: dashed line); (RIGHT) Adsorption isotherm of methane on GTCB at 113 K in the high pressure region (symbols: Experimental data; solid-line: 5-Site CS model)

Figure 1 shows the CS model is as good as that of Sun et al. [14] in terms of prediction of adsorption isotherms on graphitized thermal carbon black, and most importantly it is much less expensive than the Sun et al.'s model in terms of computation time [13]. The data at 113 K of Kiselev and co-workers extends to multilayer region and it is useful to test the potential models by comparing the GCMC simulation results to the data in this higher region. We plot in Figure 1b the GCMC results and the data in linear scale to highlight this region. Again we note the adequacy of the CS 5-Site potential model in predicting the isotherm in the multi-layer region although it is seen that there is a slight over-prediction of the data in the region of second layer formation (pressures between 40 and 60 kPa). For comparison, we also plot the GCMC simulation results obtained with the 1-Site model and the results are shown as dashed line in Figure 1b. First we note that the 1-Site model also describes well the adsorption isotherm, and secondly it also over-predicts the data in the region of second layer formation although its over-prediction is shifted towards the lower pressure range.

The microscopic configuration of methane on graphitized carbon black obtained with the 5-Site model is shown in Figure 2 where we plot the local density distribution versus the distance from the surface and the angle formed between the normal of the graphite surface and the vector pointing from the carbon atom to one of the four hydrogen atoms.

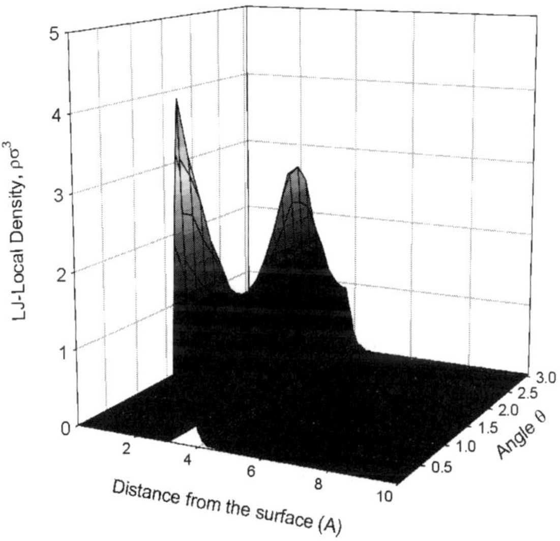

Figure 2. Local density distribution of methane. The conditions are 113 K and 1000 Pa.

An angle of zero means that the methane molecule has a pyramid configuration (tripod), while an angle of π indicates that it has an inverted pyramid configuration. Figure 2 shows that the majority of methane molecules adopts the pyramid configuration (first peak). This is physically expected because the pyramid (tripod) configuration is energetically favorable while the combination of the pyramid and inverted pyramid configurations are entropically favorable as this allows favorable packing of methane molecules having inverted tripod next to those of tripod configurations to maximize the fluid-fluid interaction.

We now finally check the potential of the 5-Site CS model by comparing the isosteric heat that is predicted from the GCMC simulation and the experimental data of Avgul and Kiselev. This is shown in Figure 3, and we see that the model predictions describe well the experimental isosteric heat, attesting to the correct 5-Site potential model in its use in adsorption studies. The isosteric heat at zero loading is correctly described, confirming the correct solid-fluid interaction, while the correct description of the linear increase of the isosteric heat with loading confirms the correct fluid-fluid interaction.

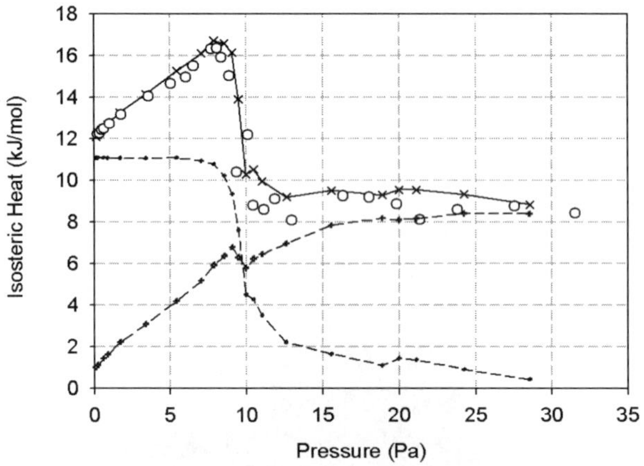

Figure 3. Isosteric heat of adsorption of methane versus loading at 113 K (circle symbols: Experimental data; solid line with cross symbols: GCMC results; dashed line with dot symbols: Heat contributed by solid-fluid interaction; dashed line with plus symbols: heat contributed by fluid-fluid interaction)

The isosteric heat can be broken down into the solid-fluid contribution and the fluid-fluid interaction. These are shown in Figure 3 as the dashed line with small dot symbols and that with small plus symbols. The heat contributed by the solid-fluid interaction is fairly constant in the region of sub-monolayer coverage (0 – 9 $\mu mol/m^2$) and this is due to the fact that most methane molecules would adopt the tripod configuration. We observe a small decline in this heat near the end of the sub-monolayer coverage and this is attributed to the fact that a small population of methane molecules adopts configurations other than the energetically favorable tripod configuration. After the first layer has been formed, the heat contributed by the solid-fluid interaction has a sharp drop and this is contributed by molecules in the second layer which is further away from the surface. The heat contributed by the fluid-fluid interaction shows a linear increase in both the first and second layers, but the rate of increase in the first layer is greater than that in the second layer.

Having described the adsorption behavior on non-porous graphitized thermal carbon black, where molecules experience no constraint in volume space for adsorption (i.e. no hindered packing effect), we would like to investigate the performance of these models for the description of methane adsorption in confined space of slit pores of various pore widths.

3.2. *Slit Pores*

The excess density in pore depends on the definition of pore volume accessible to adsorbate and therefore it is important to define this accessible volume unambiguously.

3.2.1. *Accessible Pore Volume and Width*

The accessible pore volume is defined as the volume in which a molecule can probe and the boundary of this accessible volume is defined as the loci of positions at which the solid-fluid potential is zero. If the distance from one of the pore wall to the center of the closest site of a molecule at which the solid-fluid potential is zero is z_0, the accessible pore width is

$$H' = \left(H - 2z_0 + \sigma_{ff}\right)$$

Here H is the physical width of the pore, which is defined as the distance from the plane passing through carbon atoms of the outermost layer of one wall to the corresponding plane of the opposite wall. This formula was suggested by Everett and Powl [16] and Kaneko et al. [17] for 1-Site model. For the 5-Site models, the accessible volume is calculated based on the pyramid configuration of methane because it is energetically favorable.

3.2.2. *Pore Density*

The pore density can be calculated based on the physical pore volume (AH) or the accessible pore volume (AH'):

$$\langle\rho\rangle = \frac{\langle N\rangle}{A\,H} \qquad \langle\rho\rangle' = \frac{\langle N\rangle}{A\,H'}$$

where <N> is the ensemble average of the number of particle in the simulation box, and A is the area of one wall of the pore. The plot of either <ρ> or <ρ>' versus pressure is the absolute adsorption isotherm at a given temperature, while the plot of $\langle\rho\rangle' - \rho_b$ versus pressure is the excess adsorption isotherm. It is the latter that is measured experimentally.

3.2.3. *Determination of Pore Size Distribution and External Surface Area*

The pore size distribution is denoted as f(H), with dV = f(H)dH being the physical volume of pores having physical widths falling in the range between H and H + dH. The corresponding accessible pore volume is dV' = (H'/H) f(H)dH. Therefore, the specific physical and accessible pore volumes (m^3/kg) are calculated from

$$V = \int_0^\infty f(H)\,dH\,; \quad V' = \int_0^\infty \frac{H'}{H} f(H)\,dH$$

Let ρ_{av} be the average pore density based on the physical pore volume in pores of width H. Thus, for a system containing a range of physical pore width, the number of mole in the adsorption system containing m_p (kg) of particles is:

$$N = m_p \int_0^\infty \rho_{av}(P;H)\,dV \tag{1}$$

Let us subtract and add to the RHS of eq.(1) $m_p V' \rho_b$. The result is

$$N = m_p \left[\int_0^\infty \rho_{av}(P;H)\,dV - \int_0^\infty \rho_b\,dV' \right] + m_p V' \rho_b \tag{2}$$

Rearranging the above equation we get

$$\frac{N - m_p V' \rho_b}{m_p} = \int_0^\infty \rho_{av}(P;H)\,dV - \int_0^\infty \rho_b\,dV' \tag{3}$$

The LHS of eq. (3) is the quantity that one would use to calculate the "experimental" mass excess, which is simply the difference between the amount dosed into the system and the amount that is left in the bulk phase. This quantity is a calculated one, <u>not</u> a direct experimental data. The error of this calculation would magnify greatly if the bulk gas density is comparable to the adsorbed density, which is the case at high pressures in supercritical adsorption.

The average pore density is not only a function of pressure but also on the pore width. Its dependence on pore width is significant for small pores and it becomes much less significant for pores having width greater than a threshold value H*. By splitting the integral in eq.(3) into two integrals for two different ranges of pore width, it is not difficult to obtain the following result:

$$\frac{N - m_p V' \rho_b}{m_p} = \int_0^{H^*} \left[\rho_{av}(P;H) - \rho_b \left(\frac{H'}{H} \right) \right] dV + \Gamma(P).S_{ext} \tag{4}$$

where $\Gamma(P)$ is the surface excess for surface adsorption (mol/m^2), and S_{ext} is the external surface area (m^2/kg) contributed by all pores having width greater than H* (including the outside surface area of particles).

The LHS of eq. (4) is commonly reported in the literature as the amount adsorbed (excess quantity), and this amount adsorbed when plotted against pressure is known as the isotherm commonly reported in the literature. When we use the experimental isotherm to match against the GCMC simulation results, we have to rely on the void volume, usually measured with the helium expansion method. Although it is reported that the measurements of void volume by using helium should be done at high temperatures to avoid adsorption of helium in small pores, there is no guarantee that we can eliminate completely its adsorption and resolve the situation whereby helium may access regions where adsorbate molecules can not. To avoid this uncertainty, we now introduce a new approach, which completely remove these uncertainties. This approach is outlined below.

New proposal

Since the amount introduced into the adsorption cell is accurately known, it is more convenient to report the adsorption data of supercritical conditions as the amount introduced into the adsorption cell (N) as a function of equilibrium pressure. So we rewrite eq.(4) in the following form:

$$N = m_p \left\{ \int_0^{H^*} \left[\rho_{av}(P;H) - \rho_b \left(\frac{H'}{H} \right) \right] dV + \Gamma(P).S_{ext} \right\} + m_p V' \rho_b \tag{5}$$

The significance of this equation is that the quantity required in the fitting is the amount introduced into the adsorption cell and it is always increasing with pressure. Therefore we do not have any problem with the uncertainty of the maximum in the pore density excess. Thus, such a fitting is much more reliable than the fitting of the "indirectly" calculated excess quantity versus pressure (eq. 4). So the "direct conservation of mass" equation of the form of N versus pressure will involve the pore size distribution in the range from 0 to H*, $dV = f(H)\,dH$, the external surface area of the solid (S_{ext}) and the void volume

V'. Such a determination is possible and unique solution is achievable because the average density, the surface excess and the pore density all behave differently with respect to pressure.

After the pore size distribution has been obtained, the internal geometrical area of pore walls of pores having width less than H* can be obtained as S_{int}. Thus, the total geometrical surface area is simply $S_{int} + S_{ext}$.

3.2.4. *Local Isotherms at 273 K*

Before discussing the pore size distribution of an activated carbon using adsorption of methane under supercritical conditions, we consider a number of local isotherms and discuss features associated with a number of pore sizes.

Small Micropores: 6.5A

First we show the adsorption isotherm of a very small pore (6.5 A). This pore can only accommodate one layer. Figure 4a shows the simulated absolute adsorption isotherm as well as the mass excess density isotherm using the 5-site model. The solid line with black symbols is the absolute density based on the accessible pore width, while the dashed line is that based on the physical width. The solid line is the excess density.

The two absolute densities show a monotonic increase with pressure as expected, while the excess density shows a distinct maximum, beyond which it decreases with pressure and then becomes negative at extremely high pressure. The negative relative pore density is due to the fact that the bulk density is greater than the pore density (based on accessible width), and this only occurs at extremely high pressure (~ 1000 atm). This is possible because it is easier to compress molecules in the 3D-bulk phase than in the confined space.

10 A Slit Pore

Next we show the isotherms of 10 A pore, whose width is large enough to accommodate two layers of methane molecules. Figure 4b shows the adsorption isotherms of methane over a very wide range of pressure. The behavior of the isotherms of 10 A slit pore is similar to what we have seen for smaller pores. The difference is that in this case of 10 A pore, the pore density reaches a plateau at lower pressure than those of smaller pores, and this is due to the perfect packing of two parallel integral layers of molecules in this pore. This is a direct consequence of favorable combined potential energy between solid-fluid interaction and fluid-fluid interaction.

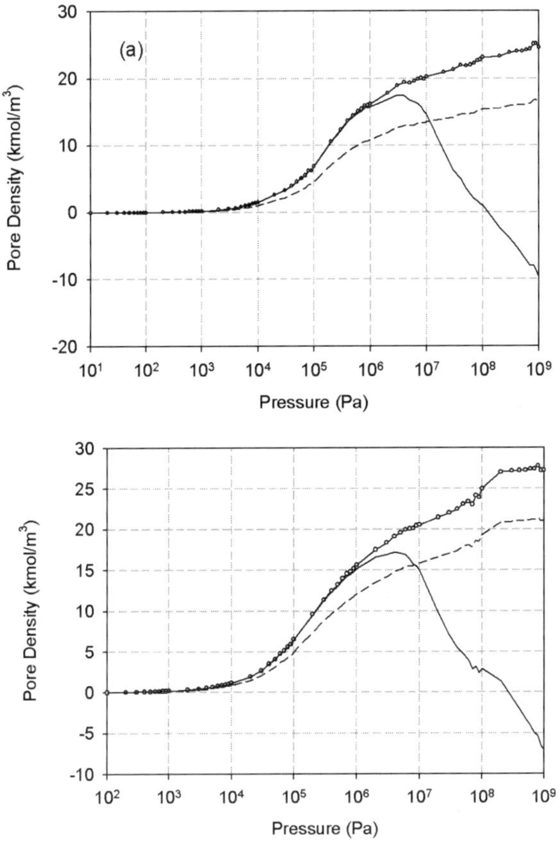

Figure 4. Isotherms of methane adsorption at 273 K (solid line with symbols is the density based on accessible width; dashed line is the density based on physical width; solid line is the excess density) (a) 6.5 A slit pore; (b) 10 A slit pore

To show the difference between the simulation result using the 5-Site model and that of 1-Site model, we observe that the absolute pore density based on accessible width using the 5-Site model is less than that using the 1-Site model (not shown). This result indicates that the 5-Site model predicts a lesser efficient packing than the 1-Site model. This observation is in agreement with the work of Boutin et al. [18] and Lachet et al. [19].

Larger pores

Adsorption in larger pores is very weak because of the weak solid-fluid interaction. A number of features that distinguish adsorption in large pores (> 20 A) to that in smaller pores are:

(1) the pressure at which the maximum of the excess density versus P is larger in larger pores

(2) the difference between the absolute density based on the physical pore width and that based on accessible pore width is getting smaller in larger pores

3.3. *Pore Size Distribution*

3.3.1. *PSD derived from 5-Site and 1-Site Methane*

The set of local excess isotherms obtained with the GCMC simulation using the 5-Site potential model for methane is produced for pore width ranging from 6.5 to 30 A. Having this set, we apply it to eq. (4) and match it against the experimental data of Zhou et al. [20] because they reported data in terms of excess density. The sample is the KOH-activated carbon and has a reported BET surface area of 3106 m^2/g and a pore volume of 1.26 cc/g. The experimental data at 273 and 233 K are shown in Figure 5 as triangle and circle symbols, respectively.

First, we use the 273 K data to fit against the theoretical isotherm to derive the PSD for this temperature. This is done by minimizing the residue which is defined as the sum of square of the differences between the theoretical and experimental isotherms. The result of this optimization is also shown in Figure 5 as the solid line using the local isotherms generated with the 5-Site potential model for methane at 273 K. We see that the fit is excellent. The pore size distribution (PSD) presented as the accessible pore volume distribution versus physical pore width is shown in Figure 6 (solid line). For this high surface area sample of KOH-activated carbon, we observe that there are two major peaks in the PSD with means of 18.5 and 26.5 A. From this distribution we derive the geometrical surface area and the pore volume as 1331 m^2/g and 1.37 cm^3/g, respectively (Table 3). The pore volume is comparable to the value of 1.3 cm^3/g, reported by the authors [20]. It is seen that the total geometric surface area obtained from this analysis is 1331 m^2/g is much lower than the BET surface area of 3106 m^2/g. It is known that the BET surface area does not represent the true geometrical area as the geometrical surface must not exceed the theoretical surface area of a single graphite layer of 2622 m^2/g, which is obtained by assuming a single layer of graphitic structure. Given the geometrical surface area of 1331 m^2/g and the theoretical area of a single graphitic layer, it could be concluded that the average number of graphite layers in this sample of high surface area activated carbon is about two.

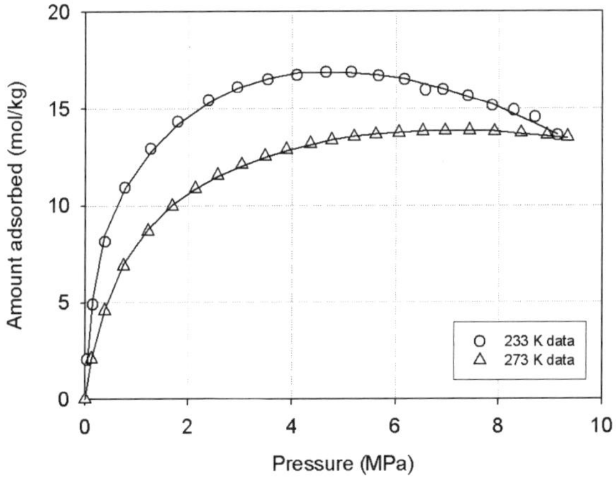

Figure 5. Experimental isotherm (symbols) of methane adsorption in high surface activated carbon. The fitted theoretical isotherm is the solid line.

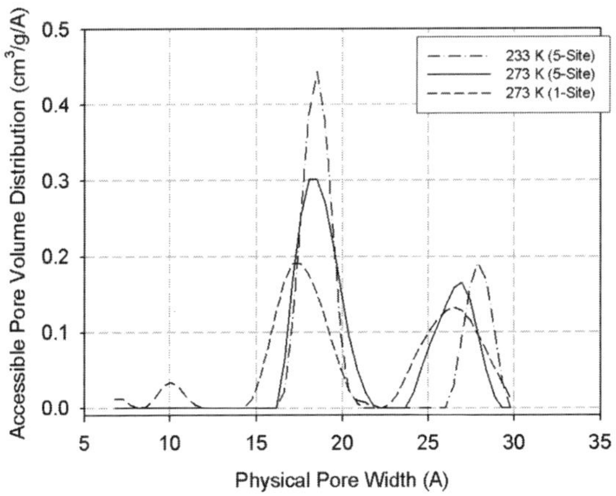

Figure 6. Accessible pore volume distribution versus the physical pore width

Table 3. Derived properties of high surface area activated carbon

	5-Site model	5-Site model	1-Site model	
	using 233 K data	using 273 K data	using 273 K data	
Accessible pore volume	1.37	1.30	1.3	cm^3/g
External surface area	~ 0	~ 0	40	m^2/g
Internal surface area	1331	1265	1353	m^2/g
Total geometric surface area	1331	1265	1393	m^2/g

Next we use the set of local isotherms obtained with the 1-Site model to obtain the pore size distribution at 273 K. The result of PSD is shown in Figure 6 as the dashed line, for which we observe that the 1-Site PSD has the first major peak shifted to the lower pore size while the second peak is quite comparable to that obtained with the 5-Site model. This is the consequence of the importance of the molecular shape of methane in small pores. The properties (pore volume, geometrical surface area) derived from the PSD-1-Site are tabulated in Table 3. Although these macroscopic properties are quite comparable to those obtained earlier with the 5-Site model, the PSD obtained with the 1-Site model is different from the one obtained earlier with the 5-Site model. In the light of the more realistic description of methane by the 5-Site model, it is expected that the results derived from this 5-Site model should be more reliable than the 1-Site counterpart model. Finally, we test another isotherm of Zhou et al. [20] at 233 K. We generate a set of local isotherms at this temperature and then derive the PSD for this temperature. The result is shown in Figure 6 as the dash-dotted line. We see that the PSD at this temperature is close to that obtained earlier at 273 K, supporting the expectation that the PSD should be temperature-independent. The macroscopic properties (pore volume and geometrical surface area) are listed in Table 3, and again we observe that they are comparable to the values obtained at 273 K.

3.3.2. *PSD derived from the new Methodology*

Now we apply the new method presented in this paper to obtain the micropore size distribution. The sample is the Ajax activated carbon. It has a BET surface area of 1200 m^2/g, and a void volume (as measured by nitrogen adsorption at 77 K) of micropores of 0.424 cc/g. Experimental data were collected with a high pressure volumetric apparatus.

Figure 7 shows the amount of methane dosed into the adsorption cell versus pressure at 273.15K. The circles denote the experimental data and the solid line is the theoretical isotherm from fitting the data using eq. (5). For clarity, the plot is given in both the linear and log-log scales.

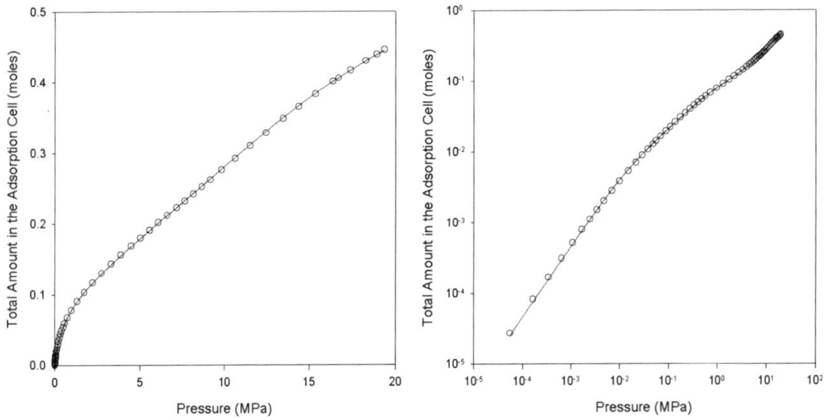

Figure 7. The amount of methane dosed into the adsorption cell versus pressure at 273.15K. Circles are the experimental data and the solid line is the fit to the data.

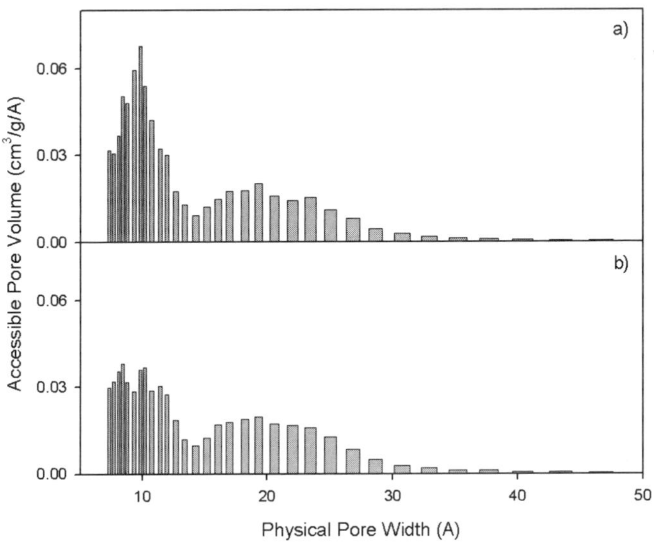

Figure 8. Pore size distributions for Ajax activated carbon from a) Fitting methane adsorption experiment data in Figure 8 and b) Using nitrogen at 77K.

The fit to the data achieved in Figure 7 is excellent. The PSD derived from this fitting is shown in Figure 8a. For comparison, the PSD derived from N_2 adsorption at 77K, used as the starting point of the PSD determination using the new method, is shown in Figure 8b. It can be seen that the two PSDs differ significantly in the range of pore sizes from 7-15A. The PSD resulting from the fit of super critical adsorption data gives a much more significant peak in the PSD in this range. Since the micropore volume of the N_2 PSD is much lower, the prediction of super critical methane adsorption using the N_2 PSD is much less than that measured experimentally. For pore sizes greater than 15A, the differences between the two PSDs are small.

The fitting of methane adsorption data was also done at 303.15K and 333.15K to test the consistency of results. The resulting PSDs from these two temperatures are given in Fig. 9.

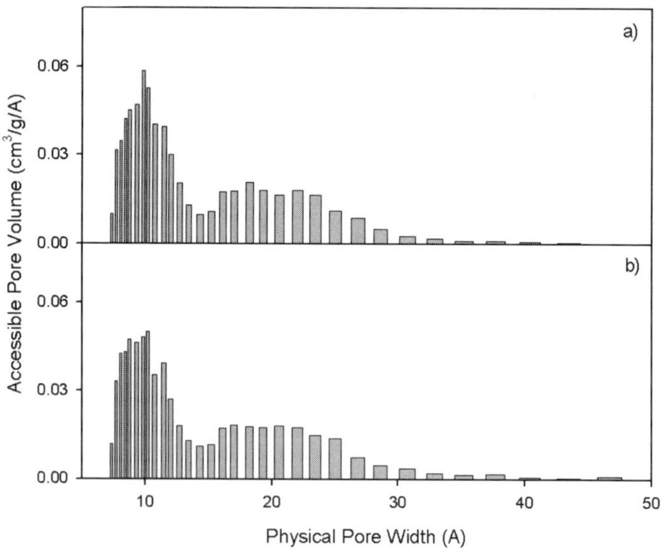

Figure 9. Pore size distributions for Ajax activated carbon from fitting methane adsorption experiment data at a) 303K and b) 333K.

The PSDs given in Figures 8a and 9 are very similar. They all have an initial peak centered about a pore size of 10A, a secondary peak at about 20A and a significantly greater micropore volume than that calculated from N_2 adsorption.

The various parameters obtained from the fitting process at different temperatures are summarized in Table 4.

So the properties given by fitting the adsorption data at different temperatures are quite consistent. The two most important things to note are the increase in the micropore volume over that measure with N_2 and the free volume of the adsorption cell being less than that measured using helium expansion. The latter is to be expected for two reasons. The first is the incidence of adsorption of helium during volume calibration and the fact that helium's smaller molecular diameter allows greater penetration into the solid than would be possible with methane. The surface areas derived from the new method represent a geometric surface are of the solid and as such, are expected to be less than the BET surface area calculated from N_2 adsorption. This is found to be the case with reasonable and comparable surface areas found at the three different temperatures. The consistency of the properties in Table 4 shows the technique to be viable.

Table 4. Properties of adsorbent from fitting experimental data as per Figure 8.

	N_2	273K	303K	333K
Micropore volume (cm^3/g)	0.424	0.471	0.465	0.462
Surface area (m^2/g)	1200	1092	1032	1055
Adsorption cell volume (cm^3)	35.1*	34.62	34.43	34.88

* from helium expansion

A further check of the new technique is to use the PSD from one temperature to predict the adsorption at 333.15 K. This is done in Figure 10 where the solid line represents the theoretical isotherm using the PSD in Figure 8a and the empty circles denote the data.

The fit in Figure 10 is very good. There is some over prediction of the amount in the adsorption cell at lower pressure (<1MPa) with the difference decreasing as the pressure increases. The source of this difference in unclear at this time and requires further investigation. The final point of interest is the difference between the excess adsorption isotherm obtained by the new method (eq. 5 less the final term for the bulk contribution to the amount in the adsorption cell) and that obtained from a traditional analysis of volumetric adsorption experiments. Excess adsorption isotherms at 273.15K and 333.15K are plotted in Figure 11.

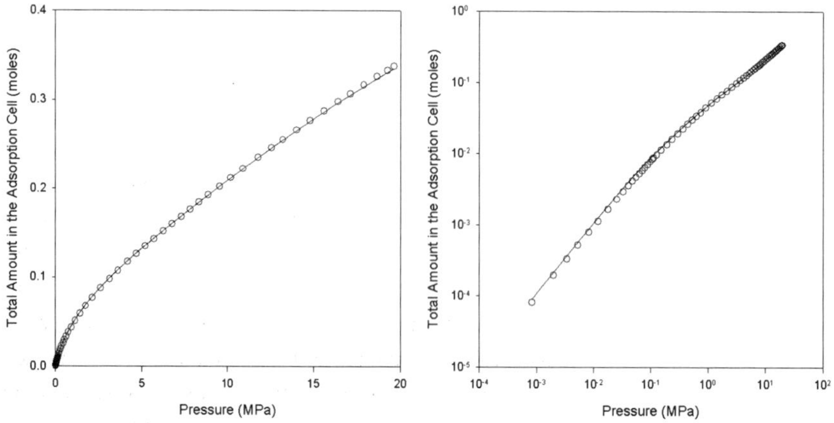

Figure 10. The amount of methane dosed into the adsorption cell versus pressure at 333.15K. Circles are the experimental data and the solid line is the theoretical isotherm using the PSD derived from data at 273K.

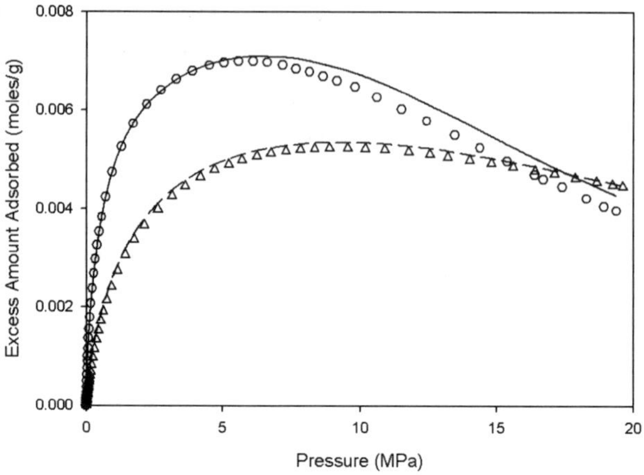

Figure 11. Excess adsorption isotherms at 273.15K (circles) and 333.15K (triangles) with the lines indicating the theoretical isotherms from the PSD in Figure 9a.

Figure 11 shows a clear difference between the isotherms obtained experimentally by a traditional treatment of the data (with the adsorption cell volume estimated by helium expansion) and the new method. The lower adsorption cell volume obtained by the new method leads to greater excess amounts adsorbed. The difference is much greater for the isotherm at 273.15K than it is at 333.15K and in the region of high pressure. Surprisingly the two methods do not diverge with increasing pressure. Instead at the highest pressures measured, the differences decrease. This is one of the many aspects of this new technique require further study. However the potential of the technique to eliminate the ambiguity of free volume measurement in high pressure adsorption is clear.

4. Conclusions

We have presented in this paper a new method to obtain the micropore size distribution using supercritical adsorption data. The potential of this method is very clear as it avoids the need to use helium expansion to determine the void volume and the uncertainty of the maximum in the mass excess versus pressure.

Acknowledgements

Support from the Australian Research Council is gratefully acknowledged.

References

1. Allen, M.P.; Tildesley, D.J. Computer Simulation of Liquids, Clarendon Press, Oxford (1987); Nicholson, D. and N. G. Parsonage, Computer Simulation and the Statistical Mechanics of Adsorption, Academic Press, London, 1982
2. Frenkel, D.; Smit, B. Understanding Molecular Simulations. Acad. Press, NY (2002).
3. El-Merraoui, M.; Aoshima, M.; Kaneko, K. Langmuir 2000 16, 4300.
4. Tanaka, H.; Merraoui, M.; Kodaira, T.; Kaneko, K. Chem. Phys. Lett. 2002 351, 417
5. Murata, K.; Kaneko, K. J. Phys. Chem. B 2001 105, 8498
6. Avgul, N.N.; Kiselev, A.V. Phys. Chem. and Phys. of Carbon 1970 6, 1; Bezus, A.G.; Dreving, V.P.; Kiselev, A.V. Russ. J. Phys. Chem. 1964 38, 1589; Isirikyan, A.; Kiselev, A. J. Phys. Chem. 1961 65, 601.
7. Gardner, L., Kruk, M., Jaroniec, M. J. Phys. Chem. B 2001 105, 12516
8. Kruk, M.; Li, Z.; Jaroniec, M.; Betz, W. Langmuir 1999 15, 1435
9. Ryckaert, J.; McDonald, I.; Klein, M. Mol. Phys. 1989 102, 2578
10. Moller, M.; Tildesley, D.; Kim, K.; Quirke, N. J. Chem. Phys. 1991 94, 8390
11. Martin, M.; Siepmann, J.I. J. Phys. Chem. B 1998 102, 2569
12. Chen, B.; Siepmann, I.J. J. Phys. Chem. B 1999 103, 5370
13. Do, D. D.; Do, H. D. J. Phys. Chem. B 2005 109, 19288
14. Sun, Y.; Spellmeyer, D.; Pearlman, D.; Kollman, P. J. Am. Chem. Soc. 1992 114, 6798
15. Steele, W. A. Surf. Sci. 1973 36, 317

16. Everett, D.; Powl, J. J. Chem. Soc. Farad. Trans. 1976 72, 619
17. Kaneko, K; Cracknell, R.; Nicholson, D. Langmuir 1994 10, 4606
18. Boutin, A.; Pellenq, R.J.M.; Nicholson, D. Chem. Phys. Lett. 1994 219, 484
19. Lachet, V.; Boutin, A.; Pellenq, R.J.M.; Nicholson, D.; Fuchs, A.H. J. Phys. Chem. 1996 100, 9006
20. Zhou, L.; Zhou, Y.; Li, M.; Chen, P.; Wang, Y. Langmuir 2000 16, 5955

ADSORPTION STUDIES OF CAGE-LIKE AND CHANNEL-LIKE ORDERED MESOPOROUS ORGANOSILICAS WITH VINYL AND MERCAPTOPROPYL SURFACE GROUPS

MIETEK JARONIEC AND RAFAL M. GRUDZIEN

Department of Chemistry, Kent State University, Kent, OH 44242, USA.

E-mail: jaroniec@.kent.edu

Ordered mesoporous cage-like silicas, FDU-1, with pendant vinyl and mercaptopropyl groups were synthesized via direct co-condensation of triethoxyvinylsilane with tetraethyl orthosilicate (TEOS), and 3-mercaptopropyl trimethoxysilane with TEOS. Moreover, vinyl-modified FDU-1 was prepared via post-synthesis modification of FDU-1 with triethoxyvinylsilane. For comparison, ordered mesoporous channel-like silica, SBA-15, with mercaptopropyl groups was synthesized by using both aforementioned methods. Nitrogen and argon adsorption-desorption isotherms provided evidence that short ligands such as vinyl can be easily incorporated to cage-like pores by both methods. The resulting materials possessed narrow pore size distributions (PSDs) and uniform openings of cage-like pores. In the case of FDU-1 with mercaptopropyl groups, argon adsorption indicated narrow PSD, whereas desorption showed nonuniformity of the pore entrance sizes. Furthermore, for the latter materials the removal of polymeric template was much more difficult.

1. Introduction

Mesoporous molecular sieves (MMSs) [1,2], which were initially prepared by self-assembly of silica precursors and ionic surfactants (alkyltrimethylammonium surfactants), are of great importance for nanoscience and nanotechnology. Few years after the discovery of MMSs [1,2] scientists started to explore the possibility to enlarge the pore size in these materials by using environmentally friendly and commercially available nonionic block copolymers as templates [2-6]. This strategy afforded MMSs of various structures, high adsorption capacity and better thermal and hydrothermal stability. One of the most popular polymer-templated ordered mesoporous silicas is SBA-15 [3,4], which was prepared by self-assembly of tetraethyl orthosilicate (TEOS) and poly(ethylene oxide)-poly(propylene oxide)-poly(ethylene oxide) triblock copolymer (PEO-PPO-PEO). SBA-15 [3,4] represents 2-D hexagonal structure (P6mm) of channel-like mesopores interconnected through small irregular pores, mostly micropores. It differs from its surfactant-templated counterpart, MCM-41, by having thicker walls, larger mesopores (up to 15 nm compared to 2-5 nm pores in MCM-41) and complementary micropores. Another important polymer-templated MMS is FDU-1 [5,7] synthesized using similar strategy but different block copolymer, which contains a more hydrophobic polybutylene oxide (PBO) block instead of polypropylene oxide (PPO). The synthesis of FDU-1 [5,7] in the presence of poly(ethylene oxide)-poly(butylene oxide)-poly(ethylene oxide) triblock copolymer (PEO-PBO-PEO) led to a 3-D cubic structure (Fm3m) with cage-like mesopores. Each spherical cage in this mesostructure is connected with twelve identical cages via short pores (apertures). Such arrangement of large cage-like mesopores and small apertures is advantageous for selective adsorption and catalysis.

A natural development in the area of MMSs was incorporation of organic functionalities [2, 8-14], which led to the so-called ordered mesoporous organosilicas (OMOs) that possess active organic ligands, also known as functional, and/or inactive organic ligands that bring additional properties apart those originated from a change in the surface polarity. The introduction of organic groups into MMSs creates tremendous opportunities for the design materials for catalysis, adsorption, sensing and so on. Currently, there are four different methods for the incorporation of organic functionalities into ordered mesoporous silicas (OMSs). The first one involves a post-synthesis modification of the template-free OMS [2, 8-9] (see panel A in Scheme 1), in which surfactant was removed by either treating nanocomposite at elevated temperatures in flowing air (calcination) or by performing extraction in acidified ethanolic solution. The second method involves the post-synthesis modification of template-containing OMS combined with simultaneous template removal [10]. Another method for creation of surface organic groups is the degradation of periodic mesoporous organosilicas (PMOs) containing bridging groups in the framework that undergo thermal reaction forming "hanging" groups on the surface (panel C). The fourth method involves one-pot synthesis (co-condensation synthesis) of desired organosilanes (see panel B in Scheme 1) [11-14].

From the practical point of view and simplicity of the synthesis procedure, a direct co-condensation [11-14] became the most prominent approach that affords ordered mesoporous materials with relatively high concentration of organic groups without losing structural ordering of the resulting material. However, post-synthesis modification [2, 8-9] permits to tailor easier the pore diameter of OMOs, which initially is governed by silica matrix (see Scheme 1A). The pore size of the starting silica depends on the nature of structure directing agent and can be tailored by varying the chain length (in the case of ionic surfactants), selecting the block copolymer of desired composition of hydrophobic and hydrophilic blocks or treating hydrothermally the self-assembled material for an extended period of time to cause its restructuring. Finally, the pore diameter can be tailored by the size and concentration of incorporated ligands. In contrast to the pore size tailoring by post-synthesis modification [2, 8-9], co-condensation [11-14] offers less possibilities to tune the pore diameter (see Scheme 1B). In the latter case the structural shrinkage is avoided because of the lack of calcination at higher temperature that substantially decreases (even up to 25%) the resulting pore width.

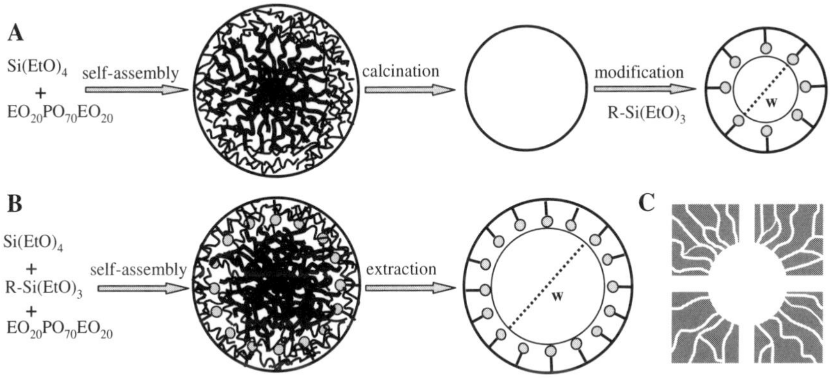

Scheme 1. Schematic illustration of incorporation of organic surface groups into mesoporous structure by two main methods: (A) post-synthesis modification (top scheme) and (B) direct co-condensation (left bottom scheme). Cage-like mesopore (C): large circle connected with straight channels represents interconnected spherical cages (ordered mesopores), whereas curved thin ribbons denote irregular micropores within walls of ordered mesopores.

One-pot synthesis [11-14], in addition to the surface modification, is widely used for the preparation of framework-modified materials known as periodic mesoporous organosilicas (PMOs) [15]. They are synthesized by hydrolysis and condensation of bis(trialkoxysilyl) organic precursors and related compounds in the presence of both ionic and nonionic templates. In contrast to the conventional OMOs that possess surface organic groups [2, 8-14], the PMO framework contains Si-R-Si linkages (where R is an organic bridging spacer) [15].

In particular, a lot of attention has been paid to OMOs with channel-like structures [1-4, 8-14] such SBA-15 [3,4,14], because these materials usually posses high adsorption quality in terms of the large pore size, high pore volume and surface area as well as high achievable loadings of surface groups.

Furthermore, the popularity of materials with channel-like pores is also due to the easiness of template removal. In contrast, the number of reports devoted to the modification of cage-like materials is limited. Their syntheses are still more challenging than those for channel-like materials because of the need to control not only the pore size distribution (PSD) but also the uniformity of the pore openings [5-7]. Also, for these materials the template removal without degradation of bridging groups is often a difficult task.

Herein, the synthesis of cage-like silicas, FDU-1 [4,5], with two different surface groups such as vinyl (V) [13] and mercaptopropyl (S) [14] is discussed. These groups were incorporated into FDU-1 by direct co-condensation [11-14] or post-synthesis modification [2, 8-9]. For comparison, channel-like silica, SBA-15, was functionalized with mercaptopropyl groups by using two aforementioned methods. Furthermore, the influence of organic groups as well as the methods of their incorporation on the adsorption properties of the resulting organosilicas is discussed.

2. Materials and Methods

2.1. *Reagents*

Structure directing agents such as poly(ethylene oxide)-poly(propylene oxide)-poly(ethylene oxide) triblock copolymer Pluronic P123 ($EO_{20}PO_{70}EO_{20}$) and poly(ethylene oxide)-poly(butylene oxide)-poly(ethylene oxide) triblock copolymer B50-6600 ($EO_{39}BO_{47}EO_{39}$) were provided by BASF Corporation and Dow Chemicals, respectively. The silica source; tetraethyl orthosilicate (TEOS, 98%) was purchased from Across Organics, whereas surface groups precursors; triethoxyvinylsilane (VS, 97%) and 3-mercaptopropyl trimethoxysilane (MPS) were obtained from Across Organics and Gelest, Inc., respectively. Fuming hydrochloric acid (HCl, 37 %) and ethanol (C_2H_5OH, 95 %) were purchased from Fischer Scientific. Deionized water (DW) was obtained at 17.5 MΩ cm using in-house Ionpure Plus 150 Service Deionization ion-exchange purification system. All chemicals were used as received without further purification.

2.2. *Synthesis of cage-like FDU-1 pure and functionalized silicas*

FDU-1 [4,5] silica was synthesized from tetraethyl orthosilicate (TEOS) in the presence of poly(ethylene oxide)-*block*-poly(butylene oxide)-*block*-poly(ethylene oxide) triblock copolymer ($EO_{39}BO_{47}EO_{39}$; B50-6600) used as template in an analogous way to that reported by Yu et al. [4]. In a typical synthesis batch 2 g of triblock copolymer was dissolved in 120 ml of 2M HCl followed by addition of 8.32 g of TEOS under vigorous stirring for 6 hours at room temperature. The resulting mixture was subsequently aged at 100 °C for 6 hours under static conditions. Finally, after filtering and washing with deionized water (DW) the slurry was dried overnight, and calcined in air at 540 °C for 4 hours to remove the template. On the other hand, vinyl-functionalized and mercaptopropyl-functionalized FDU-1 silicas (vinyl and mercaptopropyl are denoted by V and S, respectively) were synthesized similarly to the FDU-1 silica [4,5] but instead of TEOS a mixture of the specified amount of organosilane such as triethoxyvinylsilane (VS) or 3-mercaptopropyl trimethoxysilane (MPS) together with TEOS was used to achieve a desired composition. The resulting samples were synthesized by direct co-condensation method (symbol o is used to denote these samples) and assigned as FDU-1, FDU-1Vo, FDU-1So, where the sample codes listed refer to the calcined silica, extracted silicas decorated with vinyl surface groups and extracted silica functionalized with mercaptopropyl surface groups. It is noteworthy that in order to remove the template, organosilicas were extracted three times with 2 ml 98% H_2SO_4 and 100 ml of 95 % EtOH at 70 °C.

2.3. *Synthesis of channel-like SBA-15 pure and functionalized silicas*

On the other hand, SBA-15 [3, 4] mesoporous silica was synthesized from TEOS [3], whereas mercaptopropyl-functionalized SBA-15 silica was synthesized by co-condensation of MPS and TEOS in the presence of poly(ethylene oxide)-poly(propylene oxide)-poly(ethylene oxide) triblock copolymer ($EO_{20}PO_{70}EO_{20}$; P123) similarly to Zhao et al. [3] procedure (for details see [3-4, 14] and references therein). In the case of pure SBA-15 silica, 4g of polymer was dissolved in 144 ml of 1.7 M HCl under stirring for 4 hrs at 40° C followed by addition of 8 g TEOS. The synthesis mixture was kept under vigorous stirring for 24 hrs followed by heating at 100°C for 48 hrs. Analogously SBA-15 with mercaptopropyl surface ligands [14] was synthesized using the specified amounts of MPS and TEOS added to the polymer solution. The white precipitates were washed with DW, filtered and dried overnight. The resulting samples are referred to as SBA-15 and SBA-15So, where the sample codes stand for calcined SBA-15 silica and extracted mercaptopropyl-functionalized SBA-15, respectively. In addition, vinyl-functionalized FDU-1 silica and mercaptopropyl-functionalized SBA-15 silica were calcined at 550 °C in flowing air for 4 hrs to remove completely organic functionality; these samples are denoted as FDU-1Vo-c and SBA-15So-c, respectively, where c refers to the calcined samples.

Also, the vinyl-grafted FDU-1 and mercaptopropyl-grafted SBA-15 were prepared by post-synthesis modification (m) of the corresponding silicas with VS and MPS, respectively, similarly to the procedure used for the post-synthesis modification reported elsewhere [2, 8-9]. The resulting samples were denoted as FDU-1Vm and SBA-15Sm, where m stands for the post-synthesis modification.

2.4. *Adsorption and elemental analysis data collection*

Argon and nitrogen adsorption isotherms were collected using ASAP 2010 and ASAP 2020 volumetric analyzers manufactured by Micromeritics, Inc. (Norcross, GA). Adsorption isotherms were measured at -196 °C over the interval of relative pressures from 10^{-6} to 0.995 using ultra high purity argon and nitrogen from Messer Mg Industries (Malver, PA, USA) and Praxair Inc. (Danbury, CT, USA), respectively. These gases were used to measure the amount adsorbed as a function of the equilibrium pressure. Prior each adsorption measurement pure and functionalized materials were outgassed under vacuum in the port of the adsorption instrument for at least 2 hours at 200 °C and 110 °C, respectively, until the residual pressure decreased to 6 or less μmHg. Temperature 110 °C was used to avoid any bond cleavage of surface groups and to evacuate adsorbed gases, ethanol and water.

Quantitative estimation of surface groups was carried out by CHNS analysis. Nitrogen and sulfur contents for all organosilicas were determined using a LECO model CHNS-932 elemental analyzer from St. Joseph, MI.

2.5. *Calculations*

The specific surface area (S_{BET}, m^2/g) for all samples was calculated from adsorption isotherms using the Brunauer-Emmett-Teller (BET) method [16] in the range of relative pressures from 0.05 to 0.2. The volume of complementary pores [17] V_c (cm^3g^{-1}) that includes irregular small pores (mainly micropores) present in the cage-like and channel-like mesopore walls as well as interconnecting ordered apertures in cage-like structures, was estimated by integration of the initial part of the pore size distribution. The single-point pore volume (V_t, cm^3g^{-1}) [17] was calculated from the amount adsorbed at a relative pressure p/p_o of 0.99, where p and p_o denote the equilibrium pressure and saturation vapor pressure, respectively. The pore size distribution (PSD) was obtained from the adsorption branch of adsorption isotherms by employing the KJS (Kruk-Jaroniec-Sayari) method [18]. It is noteworthy that this method is based on the BJH (Barrett-Joyner-Halenda) algorithm for cylindrical mesopores [19], in which an accurate statistical film thickness and the relation between the pore size and capillary condensation pressure, established for a series of MCM-41 silicas, were employed. The diameter of ordered mesopores (w_{KJS}, nm) was found at the maximum of PSD. It was shown elsewhere [20] that the KJS method tends to underestimate the mesopores of FDU-1 by about 2 nm.

3. Results and Discussion

Shown in Fig 1A and 1B are argon and nitrogen adsorption isotherms measured at -196 °C for extracted cage-like vinyl-functionalized silicas synthesized via direct co-condensation (FDU-1Vo) and via post-synthesis modification (FDU-1Vm). These figures show also argon and nitrogen adsorption isotherms measured on the calcined silica, FDU-1. In addition, nitrogen isotherm measured on the calcined vinyl-silica is presented in Fig. 1B. Adsorption parameters such as the BET specific surface area, single-point pore volume, micropore volume and mesopore diameter evaluated on the basis of these isotherms are summarized in Table 1. All adsorption isotherms are type IV, which is characteristic for mesoporous materials that possess pores in the range between 2 nm and 50 nm. The behavior of adsorption isotherms at the range of low relative pressures indicates the presence of micropores that are typical for polymer templated silica-based materials. It is noteworthy that micropores are formed by hydrophilic chains of block copolymer, which penetrate the siliceous walls of as-made materials. The corresponding pore size distributions (PSDs in Fig. 1C and Fig. 1D) evaluated by the KJS method elaborated for the cylindrical pore geometry [17] exhibit a significant amount of porosity in the range of 1-4 nm. For the FDU-1Vo sample prepared by co-condensation the aforementioned contribution is smaller than that for the purely siliceous FDU-1 material; however, it becomes even smaller for FDU-1Vm synthesized by post-synthesis modification, indicating a partial blocking of micropores by attached vinylsilyl groups.

At higher relative pressures each isotherm curve shown in Fig. 1A and Fig. 1B exhibits a steep step that reflects the capillary condensation of adsorbates in uniform mesopores. As can be seen from Fig. 1A, the FDU-1Vo and FDU-1Vm samples feature sharp condensation steps at relative pressures of about 0.75 and 0.82, respectively, suggesting high uniformity of mesopores (narrow pore size distributions - see Fig. 1C). For the FDU-1Vm sample its pore size was about 0.7 nm smaller than that for original silica (10 nm), whereas the FDU-1Vo sample exhibited the pore size of 8.2 nm, which is confirmed by a shift of the capillary condensation step towards lower relative pressures (see argon isotherm for FDU-1Vo).

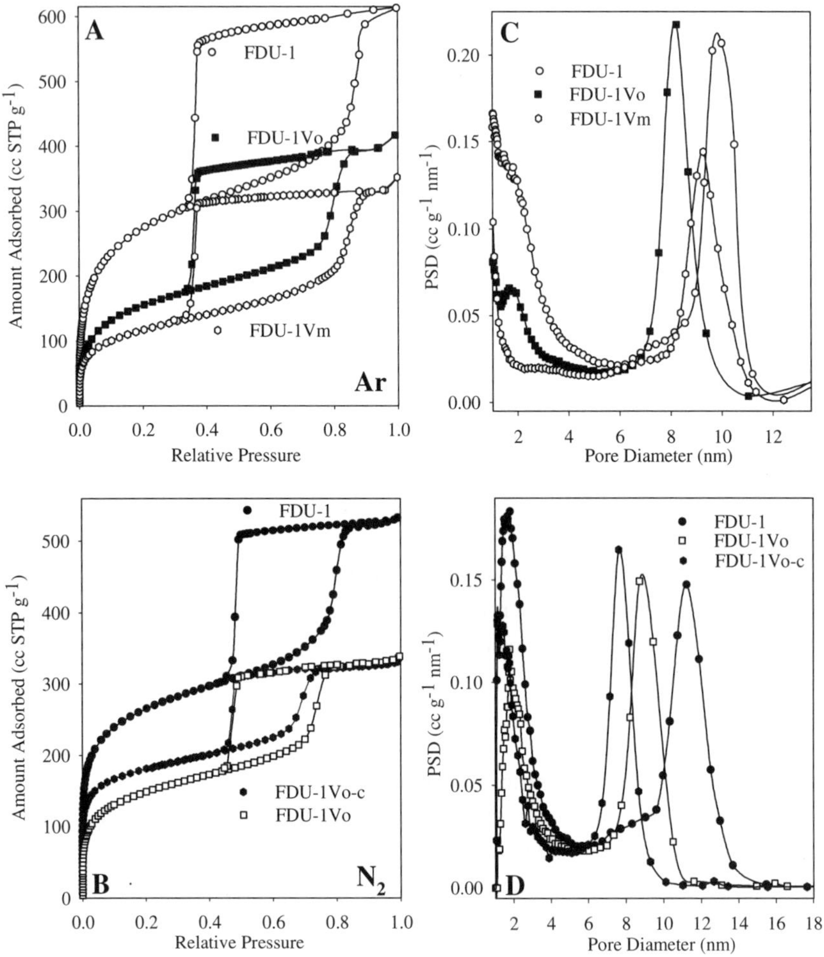

Figure 1. Comparison of argon and nitrogen adsorption-desorption isotherms measured at – 196 °C for vinyl-functionalized FDU-1 silica studied (A) and (B), respectively: calcined silica (FDU-1), extracted vinyl-functionalized silica obtained via co-condensation method (FDU-1Vo), vinyl-functionalized silica obtained via post-synthesis modification (FDU-1Vm) and calcined vinyl-functionalized silica obtained by one-pot synthesis (FDU-1Vo-c). The corresponding pore size distributions (PSDs) calculated according to the KJS method [17] from adsorption branches (C) and (D).

A visual inspection of argon and nitrogen desorption branches, which represent capillary evaporation steps, show that they are steep too, and indicate high uniformity of the pore entrance sizes. Adsorption and desorption branches of an isotherm may not coincide, which results in adsorption hysteresis loop as in the case of the samples studied. For the adsorption systems studied the observed hysteresis loops close at the limiting values of relative pressures (about 0.35 for argon at -196 °C and about 0.45 for nitrogen at -196 °C), which is characteristic for the cage-like materials with relatively small cage entrances. In the case of argon at -196 °C (Fig. 1A), there is an additional advantage because its hysteresis closes at lower relative pressure that increases the range of the pore entrance size assessment about 1 nm in comparison to that offered by nitrogen. Since for argon at -196 °C the lower limit of the pore entrance size assessment is about 4 nm and since the hysteresis loops close at the limiting relative pressure, the size of the pore openings for the vinyl-silicas studied should be not greater than 4 nm. To investigate whether adsorption properties change after removal of surface groups, the sample with vinyl groups synthesized by co-condensation was calcined at 550 °C in air. Analysis of nitrogen adsorption isotherm for this sample (Fig. 1B) shows that a complete removal of vinyl functionality reduced the mesopore diameter from 8.7 nm to 7.6 nm (see PSD in Fig. 1D) but retained its ordered porous structure.

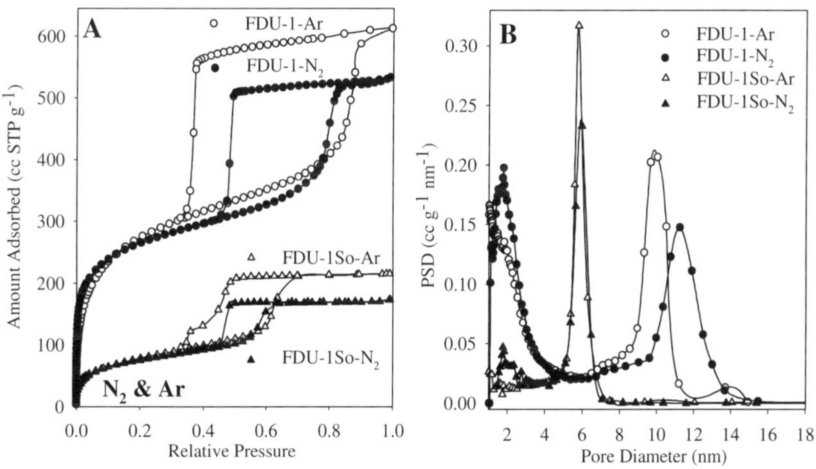

Figure 2. Comparison of argon and nitrogen adsorption-desorption isotherms measured at -196 °C for the mercaptopropyl-functionalized FDU-1 silica studied; (A): calcined silica (FDU-1) and extracted mercaptopropyl-functionalized sample obtained by co-condensation synthesis (FDU-1So), and (B) the corresponding pore size distributions (PSDs) calculated according to the KJS method [17] from adsorption branches.

In the case of mercaptopropyl-functionalized FDU-1 silica synthesized by co-condensation method (FDU-1So), argon adsorption isotherm (see Fig. 2A) exhibits sharp condensation branch that reflects uniform pore size of 5.8 nm with narrow PSD (Fig. 2B). However, argon desorption branch shows a broad step indicating non-uniformity of the pore entrance sizes, which is not seen on the corresponding nitrogen desorption branch. A comparison of the FDU-1Vo and FDU-1So samples obtained by co-condensation synthesis and having analogous concentration of surface groups indicates a significant reduction for the latter in the BET surface area from 534 m^2g^{-1} to 271 m^2g^{-1} and the total pore volume from 0.52 cm^3g^{-1} to 0.27 cm^3g^{-1}, which is mainly caused by larger ligand size, mercaptopropyl vs. vinyl. Although the post-synthesis modification with mercaptopropyl ligands was not performed for the FDU-1 silica, it is believed that the incorporation of these groups into mesoporous cages via small apertures would be difficult and could cause the pore blocking.

Mercaptopropyl-functionalized SBA-15 silicas synthesized by both methods exhibit type IV [16] adsorption-desorption isotherms (see Fig. 3A) with steep capillary condensation/evaporation branches. The observed hysteresis loops are characteristic for channel-like pores. As can be seen from Fig. 3A, similarly as in the case of vinyl-functionalized FDU-1 samples synthesized by post-synthesis modification, the pores sizes (see PSDs in Fig. 3B) of the SBA-15 silicas with mercaptopropyl groups are larger as compared to the mercaptopropyl-functionalized silica synthesized by co-condensation method. However, the values of the BET surface area and total pore volume for mercaptopropyl-silica prepared by direct synthesis are much higher, e.g., the BET surface area for SBA-15So and SBA-15Sm was 674 m^2g^{-1} and 451 m^2g^{-1}, whereas the total pore volume was 0.85 cm^3g^{-1} and 0.63 cm^3g^{-1}, respectively. Analogously to the calcined vinyl-functionalized silica, the calcined mercaptopropyl-functionalized silica (SBA-15So) exhibited the same behavior, i.e., its ordered structure was preserved but the mesopore diameter was reduced from 6.3 nm to 5.8 nm (see Fig. 3B and Table 1); however, after removal of surface groups the BET surface area increased substantially from 674 m^2g^{-1} to 1027 m^2g^{-1}, which was also observed for calcined vinyl sample (FDU-1Vo-c). For channel-like structures such as that with mercaptopropyl groups both co-condensation and post-synthesis modification methods are suitable for achieving materials with relatively high loadings of various organic ligands, uniform mesopore sizes, large total pore volume and high surface area (see Table 1 and Fig. 3A). Functionalization of cubic structures that contain cage-like mesopores with narrow apertures is much more complicated. Thus, in this case the co-condensation method is better suited for achieving higher loading of organic ligands and for tailoring surface properties of these materials.

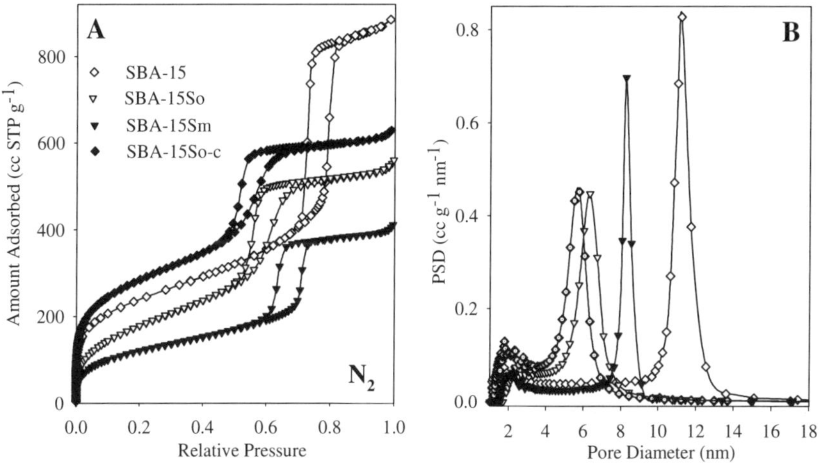

Figure 3. Comparison of nitrogen adsorption-desorption isotherms measured at -196 °C for the mercaptopropyl-functionalized SBA-15 silica studied; (A): calcined silica (SBA-15) and extracted mercaptopropyl-functionalized silica obtained by co-condensation synthesis (SBA-15So), mercaptopropyl-functionalized silica obtained by post-synthesis modification (SBA-15Sm) and calcined mercaptopropyl-functionalized silica (SBA-15So-c), and (B) the corresponding pore size distributions (PSDs) calculated according to the KJS method [17] from adsorption branches.

The incorporation of vinyl and mercaptopropyl groups to the cage-like and channel-like structures of silica was monitored by elemental analysis. The carbon (P_C) and sulfur (P_S) percentages for the samples studied are shown in Table 1. These aforementioned percentages are close to those predicted on the basis of the synthesis gel composition, which indicates an efficient functionalization of the samples studied. However, in the case of cage-like vinyl-functionalized sample a higher percentage of carbon suggests an incomplete template removal, even though this as-made sample was extracted four times with acidified ethanolic solution.

Table 1. Adsorption parameters calculated from argon and nitrogen adsorption isotherms measured at – 196 °C for vinyl-functionalized and mercaptopropyl-functionalized silicas prepared via co-condensation and post-synthesis modification.[a]

Sample	Gas	S_{BET}	V_t	V_c	w_{KJS}	P_C or (P_S)
		m^2 /g	cc/g	cc/g	nm	
FDU-1	Ar	851	0.78	0.28	10.0	0.0
	N_2	934	0.82	0.30	11.2	
FDU-1Vo	Ar	483	0.53	0.13	8.2	17.0
	N_2	534	0.52	0.16	8.7	
FDU-1Vo-c	N_2	633	0.51	0.17	7.6	0.0
FDU-1Vm	Ar	361	0.44	0.09	9.3	7.5
FDU-1So	Ar	247	0.27	0.04	5.8	(7.2)
	N_2	271	0.27	0.06	5.9	
SBA-15	N_2	855	1.36	0.14	11.2	0.0
SBA-15So	N_2	674	0.85	0.13	6.3	(3.4)
	Ar	567	0.82	0.09	6.1	
SBA-15So-c	N_2	1027	0.96	0.24	5.8	0.0
SBA-15Sm	N_2	451	0.63	0.09	8.2	(5.4)

[a]Notation: S_{BET}, BET specific surface area [16]; V_t, single-point pore volume; V_c, volume of micropores and interconnecting pores of the diameter below 4 nm; w_{KJS}, mesopore cage diameter [17]; P_C and (P_S), carbon and sulfur percentages, respectively.

4. Conclusions

Cage-like FDU-1 silicas with pendant vinyl groups, prepared via post-synthesis modification as well as co-condensation of tetraethyl orthosilicate and triethoxyvinylsilane using B50-6600 triblock copolymer as template, exhibited narrow pore size distributions and uniform pore entrance sizes. However, in the case of cage-like mercaptopropyl-functionalized silica prepared by co-condensation of 3-mercaptopropyl trimethoxysilane and tetraethyl orthosilicate, the resulting material displayed narrow PSD with nonuniform pore entrances. Moreover, mercaptopropyl-functionalized silica (FDU-1So) showed lower BET surface area, smaller pore volume and mesopore size in comparison to the vinyl-functionalized samples. In order to improve adsorption properties of cage-like ordered mesoporous silicas functionalized with organic groups (as reported recently for FDU 1 [7]) the use of lower acid concentration and addition of inorganic salt could be helpful not only to synthesize these organosilicas with larger pores and higher ligand loadings but also could be beneficial for the removal of polymeric template.

Acknowledgements

M.J. acknowledges support from the National Science Foundation Grants CTS-0553014 and CHE-0093707. The authors also acknowledge BASF Company and Dow Chemicals for providing triblock copolymers and would like to thank Kamil Gierszal from Kent State University for performing modification of SBA-15 silica.

References

1. Kresge C. T., Leonowicz M. E., Roth W. J., Vartuli J. C. and Beck J. S., Ordered mesoporous molecular sieves synthesized by a liquid-crystal template mechanism, *Nature* **359** (1992) pp. 710-712.
2. Beck J. S., Vartuli J. C., Roth W. J., Leonowicz M. E., Kresge C. T., Schmitt K. D., Chu C. T.-W., Olson D. H., Sheppard E. W., McCullen S. B., Higgins J. B. and Schlenker J. L., A new family of mesoporous molecular sieves prepared with liquid crystal templates, *J. Am. Chem. Soc.* **114** (1992) pp. 10834-10843.
3. Zhao D., Feng J., Huo Q., Melosh N., Fredrickson G. H., Chmelka B. F. and Stucky G. D., Triblock Copolymer Syntheses of Mesoporous Silica with Periodic 50 to 300 Angstrom Pores, *Science* **279** (1998) pp. 548-552.
4. Fulvio P. F., Pikus S. and Jaroniec M., Tailoring properties of SBA-15 by controlling conditions of hydrothermal synthesis, *J. Mater. Chem.* **15** (2005) pp. 5049-5053.
5. Yu C., Yu Y. and Zhao D., Highly ordered large caged cubic mesoporous silica structures templated by triblock PEO–PBO–PEO copolymer, *Chem. Commun.* (2000) pp. 575-576.
6. Grudzien R. M. and Jaroniec M., Influence of synthesis time on adsorption properties of FDU1 materials, *Stud. Surface. Sci. Catal.* **156** (2005) pp. 105-112.
7. Grudzien R. M. and Jaroniec M., Cage-like ordered silica with large mesopore volume synthesized by doubling amount of polymer, adding sodium chloride and lowering acic concentration, *Chem. Commun.* **8** (2005) pp. 1076-1078.
8. Jaroniec C. P., Kruk M., Jaroniec M. and Sayari A., Tailoring surface and structural properties of MCM-41 silicas by bonding organosilanes, *J. Phys. Chem. B* **102** (1998) pp. 5503-5510.

9. Antochshuk V. and Jaroniec M., Adsorption, thermogravimetric and NMR studies of FSM-16 materials functionalized with alkylmonochlorosilanes, *J. Phys. Chem. B* **103** (1999) pp. 6252-6261.
10. Antochshuk V. and Jaroniec M., Simultaneous modification of mesopores and extraction of template molecules from MCM-41 with trialkylchlorosilanes. *Chem. Commun.* (1999) pp. 2373-2374.
11. Burkett S. L., Sims S. D. and Mann S., Synthesis of hybrid inorganic-organic mesoporous silica by co-condensation of siloxane and organosiloxane precursors, *Chem. Commun.* (1996) pp. 1367-1368.
12. Lim M. H. and Stein A., Comparative studies of grafting and direct syntheses of inorganic-organic hybrid mesoporous materials, *Chem. Mater.* **11** (1999) pp. 3285-3295.
13. Lim M. H., Blanford C. F. and Stein A., Synthesis and characterization of a reactive vinyl-functionalized MCM-41; Probing the internal pore structure by a bromination reaction, *J. Am. Chem. Soc.* **119** (1997) pp. 4090-4091.
14. Hodgkins R. P., Garcia-Bennett A. E. and Wright P. A., Structure and morphology of propylthiol-functionalised mesoporous silicas templated by non-ionic triblock copolymers, *Microporous and Mesoporous Mater.* **79** (2005) pp. 241-252.
15. Asefa T., MacLachlan M. J., Coombos N. And Ozin G. A., Periodic mesoporous organosilicas with organic groups inside the channel walls, *Nature* **402** (1999) pp. 867-871.
16. Brunauer S., P.H. Emmet, E. Teller, Adsorption of Gases in Multimolecular Layers. *J. Am. Chem. Soc.* 60 (1938) pp. 309-319.
17. Sing K. S. W., Everett D. H., Haul R. A. W., Moscou L., Pierotti R. A., Rouquerol J. and Siemieniewska T., Reporting physisorption data for gas/solid interface with special reference to the determination of surface area and porosity, *Pure. Appl. Chem.* **57** (1985) pp. 603-619.
18. Kruk M., Jaroniec M. and Sayari A., Application of large pore MCM-41 molecular sieves to improve the pore size analysis using nitrogen adsorption measurements, *Langmuir* **13** (1997) pp. 6267-6273.
19. Barrett E. P., Joyner L. G. and Halenda P. P., The determination of pore volume and area distributions in porous substances. I. Computations from nitrogen isotherms, *J. Am. Chem. Soc.* **73** (1951) pp. 373-380.
20. Matos J. R., Kruk M., Mercuri L. P., Jaroniec M., Zhao L., Kamiyama T., Terasaki O., Pinnavaia T. J. and Liu Y., Ordered mesoporous silica with large cage-like pores: structural identification and pore connectivity design by controlling the synthesis temperature and time, *J. Am. Chem. Soc.* **125** (2003) pp. 821-829.

ADSORPTION STUDIES OF SBA-15 MESOPOROUS SILICA WITH UREIDOPROPYL SURFACE GROUPS

BOGNA E. GRABICKA, DONALD J. KNOBLOCH, RAFAL M. GRUDZIEN AND MIETEK JARONIEC

Department of Chemistry, Kent State University, Kent, OH 44242, USA.

E-mail: jaroniec@kent.edu

Ordered mesoporous organosilicas with channel-like structures (SBA15) was decorated with ureidopropyl ligands by co-condensation of ureidopropyltrimethoxysilane (UPS) and tetraethyl orthosilicate (TEOS) under high acid concentrations without addition of sodium chloride. It is shown that the co-condensation synthesis is suitable to introduce a relatively high concentration of functional ligands on the surface of channel-like mesostructures without losing their ordering, as confirmed by elemental analysis and powder X-ray diffraction (XRD). Nitrogen adsorption isotherms and pore size analysis demonstrated that the resulting mesoporous organosilicas are of high surface area, large pore volume and pore diameter in the range of 8-9 nm.

1. Introduction

The discovery of ordered mesoporous silicas (OMSs) [1-7] opened new possibilities in the area of functionalized materials [2, 8-18], which can be synthesized using commercially available functional organosilanes in the presence of structure directing agents such as ionic surfactants [2, 8, 12-17], neutral surfactants [18] and non-ionic block copolymers [9-11]. These organic-inorganic hybrids have gained growing popularity because of their potential applications in adsorption, catalysis, chromatography and host-guest chemistry for immobilization of biomolecules [9,19-21].

Frequently, functionalization of OMS is carried out to achieve the desired surface properties of the resulting material without significant changes in the specific surface area, pore volume, pore size and structural ordering. There are three major methods used to tailor the surface properties of OMSs: (i) post-synthesis grafting of the template-free OMS by using reactive organosilanes [2, 12, 15], e.g., $(C_2H_5O)_3$-Si-R, (ii) reaction of the template-containing OMS with organosilanes, which leads to the removal of the template and chemical attachment of desired surface groups [13,14], and (iii) direct co-condensation of reactive organosilanes [8-11,16-18], e.g.,

$(C_2H_5O)_3$-Si-R, and tetraethyl orthosilicate, TEOS, in the presence of structure directing agents. The latter method has been shown to be very attractive for functionalization of OMSs because it permits simultaneously to control the pore structure and to tailor the surface properties as well as to incorporate relatively high concentration of pendant groups.

In this study, we report the co-condensation synthesis of hexagonally ordered organosilica, SBA-15, with ureidopropyl (UP) surface groups on the pore walls (see scheme 1).

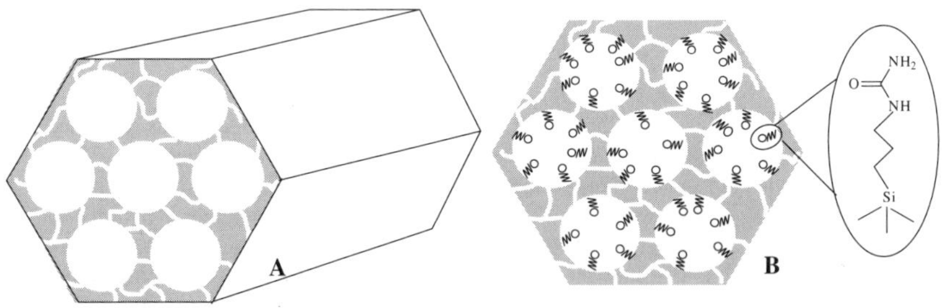

Scheme 1. Schematic illustration of hexagonally arranged channel-like mesopores in SBA-15 silica (A) and interconnected cylindrical channels (large circle with thin channels) containing ureidopropyl surface ligands (B).

2. Materials and Methods

2.1. *Reagents*

Triblock copolymer poly(ethylene oxide)-poly(propylene oxide)-poly(ethylene oxide) Pluronic P123 ($EO_{20}PO_{70}EO_{20}$) used as a structure directing agent was received from BASF Corporation. Silica source: tetraethyl orthosilicate (TEOS) was purchased from Across Organics (98 %), whereas ureidopropyltrimethoxysilane was obtained from Gelest, Inc. Deionized water (DW; conductivity < 17.5 MΩ cm) was obtained using in-house Ionpure Plus 150 Service Deionization ion-exchange purification system. Fuming hydrochloric acid (HCl, 37 %) and ethanol (C_2H_5OH, 95 %) were purchased from Fischer Scientific. All reagents were used as received without further purification.

2.2. *Synthesis of channel-like ureidopropyl-functionalized SBA-15 silicas*

Ordered mesoporous silicas, SBA-15, with ureidopropyl ligands were prepared by co-condensation synthesis of ureidopropyltrimethoxysilane (UPS) and tetraethyl orthosilicate in the presence of poly(ethylene oxide)-poly(propylene oxide)-poly(ethylene oxide) triblock copolymer ($EO_{20}PO_{70}EO_{20}$; P123) used as a structure directing agent. The synthesis recipe was similar to that reported by Zhao et al. [4]. In a typical synthesis, 2 g of polymer was dissolved in 72 ml of 1.7 M HCl under vigorous stirring at 40° C for 4-12 hours. After that a specified amount of TEOS was pipetted drop wise followed by addition of UPS to achieve the desired molar composition (see Table 1). Each solution was stirred at 40 °C for 24 h followed by hydrothermal treatment at 100 °C for 48 h. The product was filtered, washed with deionized water (DW), and dried in the oven at 80 °C. Materials were extracted three times with 2 ml HCl and 100 ml of 95 % EtOH at 70 °C to remove the polymeric template. The resulting samples are denoted as UP-*m*, where UP and *m* stand for ureidopropyl ligand and the molar percentage of incorporated surface groups, respectively. UP-*mt* denotes the as-synthesized organic-functionalized silica. The pure channel-like silica subjected to calcination at 550 °C in flowing air for 4 hours was denoted as UP-0.

2.3. *Measurements*

Nitrogen adsorption measurements were carried out using ASAP 2010 volumetric analyzers manufactured by Micromeritics, Inc. (Norcross, GA). Adsorption isotherms were measured at -196 °C over the interval of relative pressures from 10^{-6} to 0.995 using ultra high purity nitrogen from Praxair Distribution Company (Danbury, CT, USA). Nitrogen was used to measure the amount adsorbed as a function of the equilibrium pressure. All ureidopropyl-functionalized silicas were outgassed under vacuum in the port of the adsorption instrument for at least 2 hours at 110 °C prior to each measurement until the residual pressure dropped to 6 or less μmHg. Such temperature was chosen on the basis of thermogravimetric analysis to avoid the degradation of surface ligands and to remove adsorbed gases, ethanol and water.

Quantitative estimation of ureidopropyl groups was performed by CHNS analysis. Nitrogen content for all organosilicas was determined using a LECO model CHNS-932 elemental analyzer from St. Joseph, MI.

Thermogravimetric measurements were performed under flowing nitrogen on a TA Instruments Inc. (New Castle, DE, USA) model TGA 2950 high-resolution thermogravimetric analyzer. The weight change (TG) patterns were recorded over a temperature range from 35 to 800 °C. The instrument was equipped with an open platinum pan and an automatically programmed temperature controller. The high-resolution mode was used to record the TG data. The heating rate was adjusted automatically during measurements to achieve the best resolution; its maximum was 5 °C min^{-1}. The resolution and sensitivity parameters were 4 and 6, respectively. The flow rate of nitrogen gas in the system was 100 and 60 cm^3 min^{-1} on the furnace and balance, respectively.

Powder X-ray diffraction (XRD) measurements were recorded using a PANanalytical, Inc. X'Pert Pro (MPD) Multi Purpose Diffractometer with Cu $K\alpha$ radiation, operating voltage of 40 kV, 0.01° step size and 20 s step time over a range $0.5° < 2\theta < 3.0°$ at room temperature.

2.4. *Calculations*

The Brunauer-Emmett-Teller (BET) method [22] was used to evaluate the specific surface area (S_{BET}, m^2/g) in the range of relative pressures from 0.05 to 0.2 for all ureidopropyl-functionalized SBA-15 silicas. The volume of complementary pores V_c (cm^3/g) was calculated by integration the pore size distributions (PSDs) below 4 nm [23]. It is noteworthy that the volume of complementary pores contains the volume of irregular micropores present in the channel-like walls as well as the volume of small mesopores. The single-point pore volume (V_t, cm^3/g) was estimated from the amount adsorbed at a relative pressure p/p_o of 0.99, where p and p_o stand for the equilibrium pressure and saturation vapor pressure, respectively [23]. The pore size distributions were calculated from the adsorption branch of nitrogen adsorption isotherms using the KJS (Kruk, Jaroniec and Sayari) method [24], which employs the BJH (Barrett, Joyner and Halenda) algorithm for cylindrical mesopores [25] with incorporated statistical film thickness and the relation between the pore diameter and the capillary condensation pressure established for MCM-41 materials. The diameter (w_{KJS}, nm) of ordered mesopores was defined at the maximum of PSD. The primary mesopore size was also calculated by using the geometrical relation between the pore diameter (w_d, nm), volume of primary mesopores (V_p, cm^3/g), volume of complementary pores (V_c, cm^3/g), and unit cell (a, nm) derived for the *P6mm* symmetry group [26]. This relation (Equation 1) utilizes data from XRD (unit cell parameter) and gas adsorption (pore volumes) to estimate the width of ordered (primary) mesopores.

$$w_d = 1.05 \cdot a \cdot \left(\frac{V_p}{1/\rho + V_c + V_p} \right)^{1/2} \qquad (1)$$

where ρ denotes the organosilica density, which was assumed to be 2.0 g/cm^3. The wall thickness (b, nm) for hexagonal arrangement of cylindrical mesopores was calculated using Equation 2.

$$b = (a - w_d) \tag{2}$$

The unit-cell parameter (a, nm) for SBA-15 (equation 3) was evaluated using the interplanar spacing (d, nm) corresponding to (100) Bragg's reflection assessed from the X-ray diffraction profile.

$$a = d_{100} \cdot 2 \cdot 3^{-1/2} \tag{3}$$

The surface coverage of ureidopropyl ligands expressed per gram of the entire sample was estimated based on the nitrogen percentage obtained from elemental analysis.

3. Results and Discussion

The structural information for the samples listed in Table 1 was obtained from the powder XRD data, which are shown in Fig. 1. The unit cell parameters are listed in Table 2. As can be seen from Fig. 1, at least three reflections are present for the samples up to 15 %, which are indexed as (100), (110) and (200) according to the *P6mm* symmetry group. An increase in the ligand concentration to 20% caused a significant reduction of major and minor peak intensities, which indicates deterioration of the structure ordering.

Table 1. Molar composition and N% for the synthesis gels used and the corresponding N% for the SBA-15 silicas with ureidopropyl surface groups.[a]

Sample	Synthesis gel composition				Elemental analysis	
	n_{TEOS} mmol	n_U mmol	N %	C_U* mmol/g	N %	C_U mmol/g
SBA15	19.20	0	0	0	0	0
SBA15-UP5	18.24	0.96	2.16	0.77	1.20	0.43
SBA15-UP10	17.28	1.92	4.04	1.44	2.22	0.79
SBA15-UP15	16.32	2.88	5.67	2.03	3.55	1.27
SBA15-UP20	15.36	3.84	7.12	2.54	2.57	0.92

[a] n_{TEOS}, number of mmoles of TEOS; n_U, number of mmoles of UPS; C_U*, concentration of ureidopropyl groups predicted on the basis of N% in the synthesis gel mixture; C_U, concentration of ureidopropyl groups in the resulting material calculated on the basis of N% obtained by elemental analysis; % N, nitrogen percentage.

Table 2. Adsorption, structural and TG weight loss data for the samples studied.[a]

Sample	S_{BET} m^2/g	V_c cc/g	V_t cc/g	w nm	w_d nm	b nm	a nm	TG %
SBA15	866	0.14	1.38	11.2	10.4	1.10	11.50	2.88
SBA15-UP5	702	0.12	1.00	9.10	9.80	1.10	11.16	13.92
SBA15-UP10	731	0.14	1.00	9.10	10.20	1.70	11.91	15.58
SBA15-UP15	670	0.17	0.87	8.90	7.20	2.00	11.73	19.97
SBA15-UP20	525	0.16	0.42	5.8	7.2	3.4	10.60	22.11

[a] S_{BET}, BET specific surface area; V_c, volume of small pores with diameter below 4 nm obtained by integration of the PSD curve; V_t, single-point pore volume; w, mesopore diameter calculated by the KJS method [24]; w_d, mesopore diameter calculated on the basis of the unit cell parameter and pore volumes according to the relation derived for the P6mm structure [26] assuming 2.0 g/cm^3 density of silica; b, pore wall thickness; a, unit cell parameter obtained on the basis of XRD patterns; TG, thermogravimetric weight loss recorded in flowing nitrogen in the range between 100 and 800 °C.

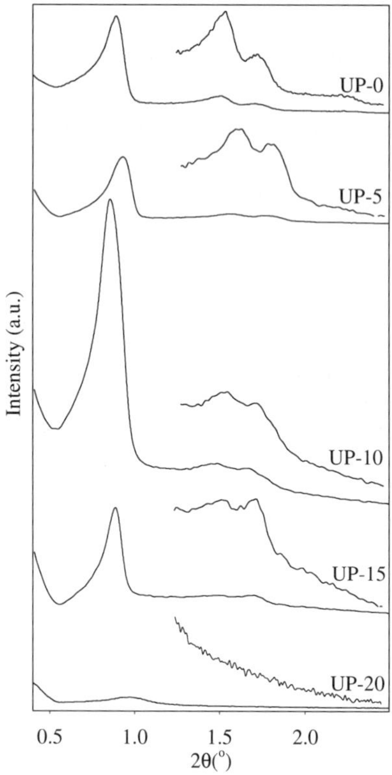

Figure 1. X-ray diffraction (XRD) patterns for the extracted mesoporous channel-like SBA-15 silicas with ureidopropyl surface groups.

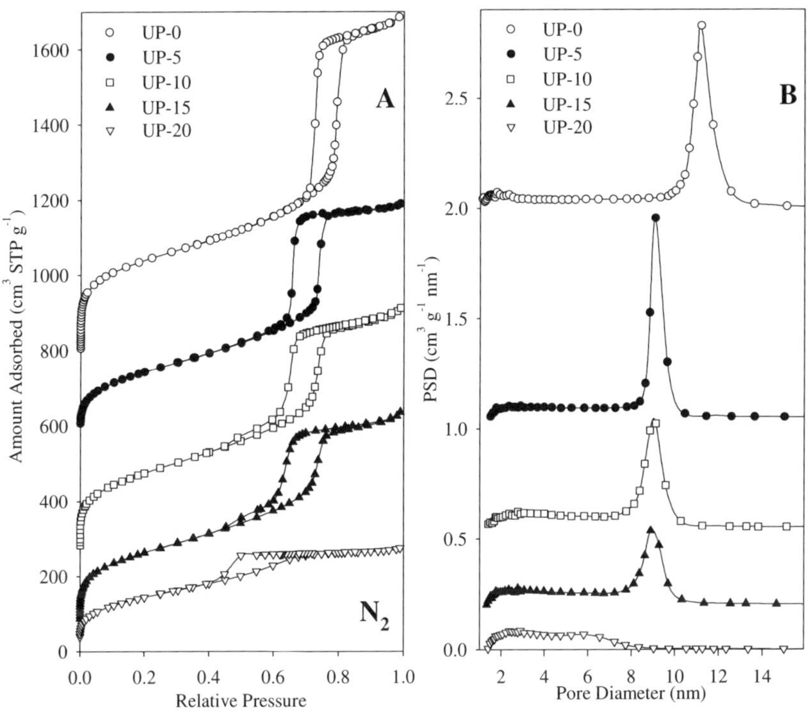

Figure 2. (A) Nitrogen adsorption-desorption isotherms measured at -196 °C for the extracted mesoporous channel-like SBA-15 silicas with ureidopropyl surface groups. The isotherms for UP-0, UP-5, UP-10 and UP-15 were offset vertically by 800, 550, 275 and 80 cc STP g^{-1}, respectively. (B) Pore size distributions (PSDs) calculated according to the KJS method [24] for each nitrogen adsorption isotherm. The pore size distributions UP-0, UP-5, UP-10 and UP-15 were shifted vertically by 2, 1.05, 0.55 and 0.2 cc g^{-1} nm^{-1}, respectively.

A comparison of nitrogen adsorption-desorption isotherms measured at – 196 °C is shown in Figure 2A. These isotherms are of type IV with sharp capillary condensation/evaporation steps and pronounced H1 hysteresis loop, which is typical for materials with cylindrical pores.

The presence of sharp capillary condensation steps on these isotherm curves (except UP-20) indicates high uniformity of pore sizes, which is reflected by narrow PSD curves (Fig. 2B). As can be seen from Fig. 2B, the PSD curves insignificantly shift to smaller pores with increasing concentration of ureidopropyl ligands. Adsorption parameters such as the BET specific surface area, volume of complementary small pores, total pore volume and mesopore diameter for the samples studied are summarized in Table 2. For instance, the sample UP-10 exhibits the BET specific surface area of 731 m^2/g, total pore volume (0.9 cc/g) and pore diameter of 9.12 nm, which are analogous to the parameters obtained for the remaining samples. However, an increase in ureidopropyl loading (UP-20) led to a meaningful PSD broadening and a decrease in the surface area and pore volume.

Figures 3A and Fig 3B show a comparison of the TG profiles recorded in nitrogen atmosphere and the corresponding DTG curves for the extracted SBA15 samples with varying percentage of ureidopropyl groups as well as for the as-made sample containing polymer template (SBA15-UP15t). As can be seen from the TG plots for SBA15-UP15t (Fig. 3A and 3B), the polymer template was completely removed after extraction, which is reflected by the disappearance of a large peak at about 375 °C, while the ureidopropyl groups remained intact as indicated by the presence of decomposition peaks in the range between 200 and 300 °C for both composite and extracted samples. The observed enlargement in the peak intensity between 200 and 300 °C with increasing percentage of ureidopropyl groups confirms a successful incorporation of this functionality.

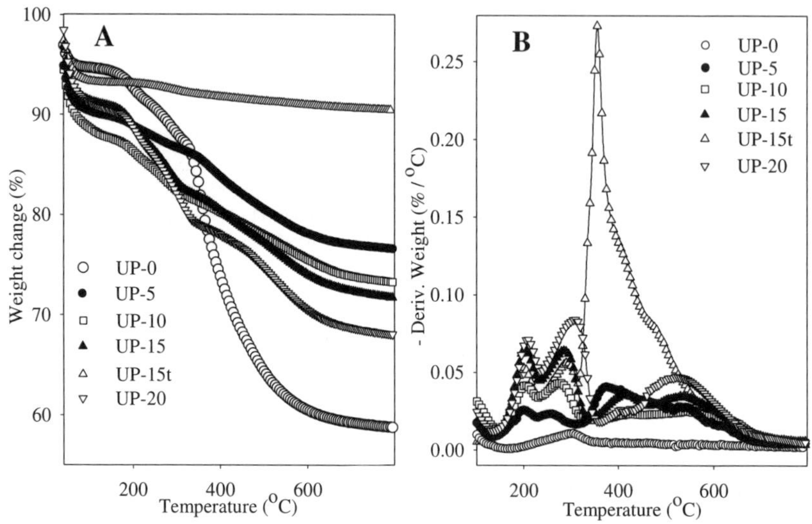

Figure 3. (A) The weight change (TG) curves measured in flowing nitrogen for the SBA-15 samples with ureidopropyl groups: calcined pure silica (UP-0) and extracted organosilicas (UP-5, UP-10, UP-15, UP-20) and as-synthesized sample (UP-15t) with various percentages of ureidopropyl ligands, and (B) the corresponding DTG curves.

The decoration of mesopore walls with ureidopropyl groups was monitored by elemental analysis (see nitrogen percentage values listed in Table 1). Nitrogen percentages obtained from elemental analysis increase with increasing amount of UPS in the synthesis gel, which suggests that the concentration of ureidopropyl groups in the resulting materials ligands increases too.

4. Conclusions

In conclusion, this work shows that the co-condensation synthesis afforded SBA-15 materials with relatively large amount of ureidopropyl groups (up to 15%) on the mesopore walls without significant deterioration of the structural ordering. The resulting materials exhibit high surface areas, large pore volume and pore widths about 9 nm.

Acknowledgements

M.J. acknowledges the National Science Foundation Grants CTS-0553014 and CHE-0093707. The authors thank BASF for providing P123 block copolymer.

References

1. Kresge C. T., Leonowicz M. E., Roth W. J., Vartuli J. C. and Beck J. S., *Nature* **359** (1992) pp. 710-712.
2. Beck J. S., Vartuli J. C., Roth W. J., Leonowicz M. E., Kresge C. T., Schmitt K. D., Chu C. T.-W., Olson D. H., Sheppard E. W., McCullen S. B., Higgins J. B. and Schlenker J. L., *J. Am. Chem. Soc.* **114** (1992) pp. 10834-10843.
3. Yanagisawa T., Shimizu T., Kuroda K. and C. Kato, *Bull. Chem. Soc. Japan* **63** (1990) pp. 988-992.
4. Zhao D., Feng J., Huo Q., Melosh N., Fredrickson G. H., Chmelka B. F. and Stucky G. D., *Science* **279** (1998) pp. 548-552.
5. Kruk M., Jaroniec M., Ko C.H. and Ryoo R., *Chem. Mater.* **12** (2000) pp. 1961-1968.
6. Fulvio P. F., Pikus S. and Jaroniec M., *J. Mater. Chem.* **15** (2005) pp. 5049-5053.
7. Zhang F., Yan Y., Yang H., Meng Y., Yu Ch., Tu B. and Zhao D., *J. Phys. Chem. B* **109** (2005) pp. 8723-8732.
8. Asefa T., MacLachlan M. J., Coombos N. and Ozin G. A., *Nature* **402** (1999) pp. 867-871.
9. Olkhovyk O. and Jaroniec M., *J. Am. Chem. Soc.* **127** (2005) pp. 60-61.
10. Grudzien, R. M., B. E. Grabicka, S. Pikus and M. Jaroniec, *Chem. Mater.* **18** (2006) pp. 1722-1725.

11. Grudzien R. M., Pikus S. and Jaroniec M., *J. Phys. Chem. B* **110** (2006) pp. 2972-2975.
12. Lim M. H. and Stein A., *Chem. Mater.* **11** (1999) pp. 3285-3295.
13. Antochshuk V. and Jaroniec M., *Chem. Commun.* (1999) pp. 2373-2374.
14. Antochshuk V. and Jaroniec M., *Chem. Mater.* **12** (2000) pp. 2496-2501.
15. Jaroniec C. P., Kruk M., Jaroniec M. and Sayari A., *J. Phys. Chem. B* **102** (1998) pp. 5503-5510.
16. Burkett S. L., Simms S. D. and Mann S., *Chem. Commun.* **11** (1996) pp. 1367-1368.
17. Kruk M., Asefa T., Coombs N., Jaroniec M. and Ozin G. A., *J. Mater. Chem,* **12** (2002) pp. 3452-3457.
18. Gong Y. J., Li Z. H., Sun Y. H., Deng F., Luo Q. and Yue Y., *Microporous and Mesoporous Mater.* **49** (2001) pp. 95-102.
19. Olkhovyk O. and Jaroniec M., *Adsorption* **11** (2005) pp. 205-214.
20. Antochshuk V., Olkhovyk O., Jaroniec M., Park I.-S. and Ryoo R., *Langmuir* **19** (2003) pp. 3031-3034.
21. Kang T., Park Y., Choi K., Lee J. S. and Yi J., *J. Mater. Chem.* **14** (2004) pp. 1043-1049.
22. Brunauer S., Emmet P.H. and Teller E., *J. Am. Chem. Soc.* 60 (1938) pp. 309-319.
23. Sing K. S. W., Everett D. H., Haul R. A. W., Moscou L., Pierotti R. A., Rouquerol J. and Siemieniewska T., *Pure. Appl. Chem.* **57** (1985) pp. 603-619.
24. Kruk M., Jaroniec M. and Sayari A., *Langmuir* **13** (1997) pp. 6267-6273.
25. Barrett E. P., Joyner L. G. and Halenda P. P., *J. Am. Chem. Soc.* **73** (1951) pp. 373-380.
26. Kruk M., Jaroniec M. and Sayari A., *Chem. Mater.* **11** (1999) pp. 492-500.

EFFECT OF POROSITY AND FUNCTIONALITY OF ACTIVATED CARBON IN ADSORPTION

FRANCISCO RODRÍGUEZ-REINOSO

Laboratorio de Materiales Avanzados. Universidad de Alicante. Apartado 99. E-03080 Alicante. Spain.
reinoso@ua.es

The presentation is concerned with the main characteristics of the well known adsorbent activated carbon, the rather high inertness of the surface, the slit-shaped microporosity, the flexibility in the porosity development and the flexibility in the modification of the chemical nature of its surface, and the effects that such characteristics have on the application of activated carbon in adsorption processes. Several examples are shown to highlight these effects, with special emphasis on the gas separation and gas storage processes.

Activated carbon is a very important industrial adsorbent because it exhibits a well developed porosity (micro, meso and macroporosity) and this is coupled with a great thermal and chemical stability, a relatively large hydrophobicity (thus favouring the adsorption of non-polar substances in the presence of humidity), low production cost, etc. Additionally, the surface of activated carbon can be functionalised with different heteroatoms (but mainly oxygen), thus modifying the chemical nature. A large and accessible surface area is a necessary but not sufficient condition for the preparation of activated carbons to be used in industrial adsorption processes (gas and liquid phase purification, separation, environmental control, etc.), since the last few years has shown that the chemical composition of the carbons surface plays a very important role in the process.

Porosity in activated carbon is rather unique since the more important range of porosity from the point of view of adsorption capacity is the microporosity, which in activated carbon is slit-shaped. This has a considerable effect on the adsorption properties of this material because: i) the microporosity can be used to separate adsorbing molecules as a function of both molecular dimension and/or shape (see Figure 1), and ii) slit-shaped microporosity is responsible for a larger packing density of adsorbed molecules relative to cylindrical-shaped pores of the same dimensions, thus facilitating the adsorption of higher amounts of gas adsorbed per unit volume of carbon (see Figure 2).

The presence of oxygen surface groups in activated carbon can completely modify the adsorption behavior of the adsorbent because in the absence of these groups the carbon surface would be rather inert and would preferably adsorb non-polar molecules. The introduction of oxygen surface groups renders the carbon surface more polar and the adsorbent will then be able to adsorb more polar substances, the uptake being an additional function of the amount of groups present. In the case of adsorption of molecules with some polarity the chemical nature of the carbons surface is very important because for instance the adsorption of water is almost nil at low relative pressures and it is not important until the pressure is high enough to produce condensation in the mesopores. However, if the carbon is slightly oxidised with hydrogen peroxide or nitric acid the shape of the isotherm drastically changes and the interaction with the water molecule becomes stronger. However, if the adsorption on the walls of the carbon porosity is taking place through the interaction of the adsorbing molecule with the π electrons of the graphene layers the presence of oxygen surface groups at the edges of these planes will withdraw electron density from the graphene layer (oxygen is highly electronegative), thus reducing the uptake of aromatic molecules such as phenols.

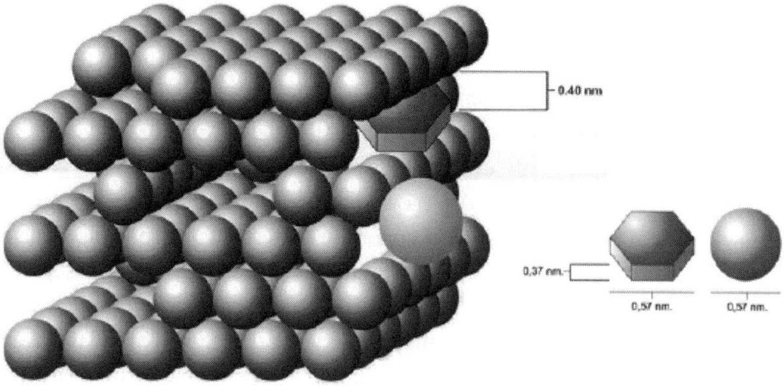

Figure 1. Model to show the selectivity for the adsorption of molecules in activated carbon.

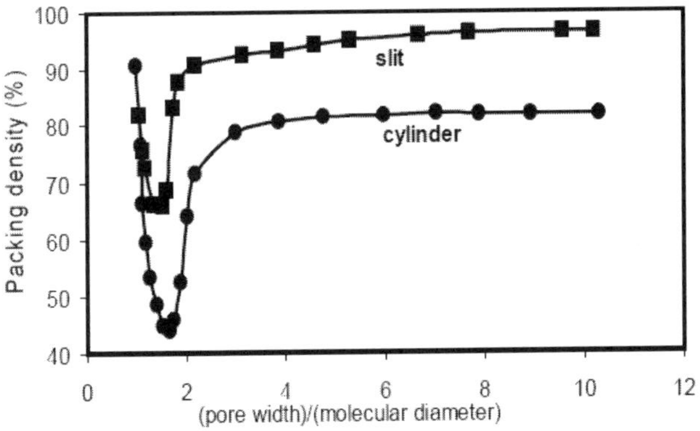

Figure 2. Packing of spherical molecules in model micropores.

In the case of gas separation, something extremely important in activated carbon is the slit-shaped microporosity, in contrast with the cylindrical porosity found in most inorganic adsorbents. This shape in the microporosity will produce a molecular sieving effect for molecules as a function of molecular dimension and shape and for this reason carbon molecular sieves are used for industrial separations. A typical example of separation based on the molecular shape is that benzene from methane, normal- from iso-parafins, etc. Additionally, separations can be based on kinetics aspects as in the case of production of nitrogen from air by pressure swing adsorption (PSA) using a 4A carbon molecular sieve because oxygen diffuses more rapidly into the microporosity, nitrogen not being adsorbed.

A derivation of activated carbon prepared for the separation of gases are the Carbon Molecular Sieves (CMS), which are more and more frequently used in industrial processes. The possible advantages of CMS in respect to conventional sieves such as zeolites for many processes are: shape selectivity for planar molecules, higher hydrophobicity, high resistance to acid and basic media and thermal stability under inert atmospheres. There are CMS which separate the components of gas mixtures on the basis of molecular size and shape. In other applications, the separation is carried out on the basis of kinetics (rates), where equilibrium adsorption uptakes, by the carbon, for both gases, are very similar. Because examples of gas separation by size exclusion are popular, as for instance the separation of benzene from cyclohexane or the separation of normal- and iso-parafins, the following information is related to the separation based on kinetics factors, examples being the preparation of nitrogen from air (the better known application of CMS) and purification of natural gas (removal of carbon dioxide).

CMS are prepared using several experimental procedures, with commercial CMS being manufactured from activated carbon by a treatment that deposits pyrolytic carbon at the entrance of the micropores until the width is reduced to the desired dimension. The main problem with this procedure is the difficulty in controlling the deposition process, which may result in a decrease of the CMS adsorption capacity.

In addition to the conventional carbon vapor deposition method, our research group has used two additional procedures for the preparation of CMS: i) controlled uncatalysed gasification of chars obtained from lignocellulosic precursors; and ii) mild oxidation of a char and subsequent controlled removal of oxygen surface groups (this second procedure can also be applied to a previous CMS with wider micropore width, to reduce the pore width).

In the first of these procedures, the lignocellulosic precursor (coconut shells or peach stones) was acid washed to eliminate mineral matter as far as possible and then slowly carbonized. The char was activated (*thermally*) with carbon dioxide at 750 °C (to ensure a slow gasification) to controlled burn-offs.

In the second procedure, the char, or a previous CMS with dimensions larger than required, was subjected to oxidation with nitric acid, which introduces significant amounts of oxygen surface groups into the char or carbon. This chemisorbed oxygen effectively reduces the entrance dimensions of the microporosity. Further fine-tuning is achieved by subsequent heat treatment under inert atmosphere to remove excess surface oxygen groups as carbon monoxide and carbon dioxide.

The porosity of CMS is studied by adsorption of N_2 (77 K) and CO_2 (273 K) to determine volumes of total and narrow microporosity, respectively, and by immersion calorimetry of the carbons into liquids with different molecular dimensions (dichloromethane, 0.33 nm; benzene, 0.37 nm; cyclohexane, 0.48nm; 2,2-dimethylbutane, 0.56 nm; and α-pinene, 0.70 nm). Adsorption kinetics were studied for two-gas mixtures, nitrogen-oxygen and methane-carbon dioxide, and separation abilities were studied using columns packed with the corresponding CMS.

Separation of nitrogen and oxygen is an optimum for two CMS prepared by CO_2 activation of the char and by nitric acid oxidation of the char and subsequent heat treatment under helium at 400 °C. The selectivity of these two CMS for this separation is 11-14, selectivity being defined as the ratio between the amounts adsorbed after 120 seconds contact with 0.1 MPa of gas. Very high values of selectivity for the CO_2/CH_4 gas separation, well above 100, were obtained in some of the CMS prepared.

In the case of gas storage (methane in the example used here) the approach is to use modifications of conventional *chemical* activation processes. From the point of view of gas storage the carbon bed can be separated into three well-defined volumes: i) carbon skeleton; ii) volume of meso- and macropores plus the volume of interparticle space (the packing density of methane would be low in this volume); and iii) the volume of micropores. A good adsorbent should exhibit a high volume of micropores and a low volume for the rest of the space, thus facilitating a high volume of gas adsorbed per unit of volume of CMS. An answer is to use monoliths of carbon in which these volumes are optimized.

The manufacture of monoliths without the need for an additional binder, by chemical activation of lignocellulosic precursors, is an interesting procedure. The generation of tars, following impregnation of the precursor with phosphoric acid or zinc chloride, under pressure, impregnates the carbon particles so stabilizing the monolith [5-7]. Further heat treatment, followed by washing of the residual chemical, leads to a final carbon artifact suitable for gas storage. However, this is not a suitable procedure for carbons obtained by chemical activation using potassium (or sodium) hydroxide of the same lignocellulosic precursor. This is because such chemical activation starts above 700°C, after the formation of the char and in this sense the activation mode for the original precursor and its char is very similar. The resultant carbon cannot be conformed under pressure without addition of a binder and, consequently, it is not adequate for gas storage.

The question is then which chemical agent is more appropriate for the preparation of activated carbon monoliths with good storage capacity for natural gas (methane in the laboratory). Zinc chloride is not very popular nowadays in the manufacture of commercial activated carbons because of the problems associated with the presence of zinc in the environment. However, it is a very interesting chemical because the activated carbons, so prepared, are dominantly microporous and, depending on the impregnation ratio used, the porosity can be extended to the lower range of mesopores, but not to larger mesopores or macropores. A typical example of these monoliths is: carbon skeleton 41%; microporosity 47% and voids (macro plus interparticle space): 12% [5].

With phosphoric acid activation, the porosity development is different because essentially microporous carbons can be prepared. However, the use of higher concentrations of this chemical also develops meso- and macroporosity. Further, a controlled process leads to carbon monoliths in which the internal volumes are as follows: carbon skeleton 38 %; microporosity 53 %, voids, 9 % [6].

The sets of monoliths prepared using both of these chemical agents can be used directly for methane storage because values higher than 100 V/V are obtained. (V/V is the ratio of gas volume to carbon volume). However, even higher values can be reached if these activated carbon monoliths are further activated by slow gasification with carbon dioxide at temperatures around 800 °C. Here, there is an enhancement of the microporosity and storage capacities around (and above) 150 V/V can be reached. This value is considered to be the lower limit of the practical application of methane storage at an industrial level. This means that careful optimization of the different steps in the manufacturing process has to be introduced in order to reach higher values. Simulation of these systems suggests that values as high as 220 V/V can be reached with microporous carbon adsorbents.

There are many examples of the effect of the chemical nature of the carbon surface on adsorption processes. In the case of activated carbons with a reduced number of oxygen surface groups the adsorption of non-polar molecules is favored and the interaction of the carbon surface with molecules such as water, methanol, etc is very reduced, leading to type III or V isotherms. However, if the carbon is oxidized with a solution of hydrogen peroxide or nitric acid, there is a large increase in the amount and variety of oxygen surface groups with a direct effect on the interaction with polar molecules, which is considerably increased. Several examples can be provided to show this effect of the chemical nature of the surface of the adsorption process, typical ones being related to the removal of volatile organic compounds (VOC) from industrial gaseous streams or the removal of phenols from water. In some cases the presence of oxygen surface groups enhances the adsorption of polar molecules but in many others the surface groups decrease the adsorption capacity. The later is the case for the adsorption of aromatic compounds such as benzene. In this case the presence of oxygen, highly electronegative, removes electronic density from the graphene layer constituting the carbon porosity, thus reducing the interaction between the π electrons of the layer with the aromatic ring of benzene and, consequently, reducing the adsorption capacity in respect to a similar carbon with no oxygen surface groups.

Acknowledgements

This work was partially funded by the Spanish MCYT (Projetc BQU2003-0615), Generalitat Valenciana (project Grupos03/212), Petrobras (Brazil) and the European Network off Excellence "Insidepores".

References

1. Marsh, H., Rodríguez-Reinoso, F. Activated Carbon. Elsevier, Amsterdam. 2006. In press
2. Gómez-de-Salazar C, Sepúlveda-Escribano A, Rodríguez-Reinoso F. Preparation of carbon molecular sieves by pyrolytic carbon deposition. Adsorption 2005; 11, 663-667.
3. Gómez-de-Salazar C, Sepúlveda-Escribano A, Rodríguez-Reinoso F. Preparation of carbon molecular sieves by controlled oxidation treatments. Carbon 2000;38(13):1889-1892.
4. Arraigada, R., Bello, G., García, R., Rodríguez-Reinoso, F., Sepúlveda-Escribano, A. Microp. Mesop. Mater. 2005; 81, 161-167.
5. De Salazar CG, Sepúlveda-Escribano A, Rodríguez-Reinoso F. Use of immersion calorimetry to evaluate the separation of carbon molecular sieves. Stud Surf Sci Catal 2000;128:303-312.
6. Almansa C, Molina-Sabio M, Rodríguez-Reinoso F. Adsorption of methane into $ZnCl_2$-activated carbon derived discs. Micropor Mesopor Mater. 2004;76(1-3):185-191.
7. Molina-Sabio M, Almansa C, Rodríguez-Reinoso F. Phosphoric acid activated carbon discs for methane adsorption. Carbon 2003;41(11):2113-2119.
8. Molina-Sabio M, Rodríguez-Reinoso F. Role of chemical activation in the development of carbon porosity. Colloids and Surfaces A: Physicochem Eng Aspects. 2004;241:15-25.
9. Rodríguez-Reinoso, F., Almansa, C., Molina-Sabio, M.; Contribution to the evaluation of density of methane adsorbed on activated carbon. J. Phys. Chem. B 2005; 109, 20227-20231.
10. Rodríguez-reinoso, F., Molina-sabio, M, Muñecas, M.A.. Effect of microporosity and oxygen surface groups of activated carbon in the adsorption of molecules of different polarity. J. Phys. Chem. 1992; 96, 2707-2713.

PHASE BEHAVIOR OF SIMPLE FLUIDS CONFINED IN COORDINATION NANOSPACE

MINORU MIYAHARA AND TAKURO KANEKO
Department of Chemical Engineering, Kyoto University, Nishikyo, Kyoto 615-8510, Japan
E-mail: miyahara@cheme.kyoto-u.ac.jp

Freezing behavior of Lennard-Jones (LJ) fluid confined in a coordination nanospace, or the metal-organic framework, was examined employing GCMC technique. A unit cell that contains at least 3x3 array of square channels divided by thin walls of single atom thickness was developed. The simulations clarified that the LJ-methane in graphene walls with the effective channel size of ca. 4σ exhibited extremely elevated freezing points. The significant elevation was considered to be brought not only by superimposed potential from walls, but also partly by the interaction between fluid molecules existing in different compartments through the ultrathin walls. Besides these factors, results of simulations with walls made up with fluid molecules themselves indicated possibility of additional enhancing factor for freezing that was not prevailing in slit-pore case.

1. Introduction

Understanding for phase behavior of confined fluids in nanospace has progressed a great deal in this decade. As for the vapor-liquid coexistence, many studies including ours have shown the incorrectness of the Kelvin model in the scale of nanometers, and an improved model for accurate pores size estimation was proposed [1]. As for the solid-liquid transition, the authors have clarified that the freezing-point temperature of confined fluid gets higher as well as lower than the bulk freezing point, which would result from combination of three factors: i) elevating effect by the pore-wall potential energy (compressing effect) [2], ii) geometrical shape of pore (geometrical hindrance effect) [3], and iii) depressing effect by the tensile condition of the capillary condensate (tensile effect) [4]. Simple thermodynamic models for solid-liquid phase boundary were proposed in the above studies.

Further research of us includes determination of triple point by molecular simulation, which can also be estimated if we take account of appropriate effects among the above [5]. Also, sublimation or gas-solid transition of LJ-methane confined in carbon nanopore has been recently examined [6]. The obtained sublimation temperature is significantly elevated, which can be predicted by a simple model with no adjustable parameter. With this success a whole Lennard-Jones phase diagram in nanopore can now be predictable. Figure 1 illustrates a typical phase diagram of simple fluid confined in slit nanospace with strongly attractive walls, superimposed on the bulk phase diagram.

Standing upon the above understanding for usual nanospace, we now, in this study, seek unique characteristics of phase behavior of simple fluids confined in nanoscale coordination space, or metal-organic framework (MOF), employing molecular simulation technique.

Uniqueness would result from ultrathin wall of the coordination space, which differs completely from usual porous materials with pore space surrounded by coarse solid phases. Thus the confined fluids may feel not only the wall-fluid interactions, but also those from fluid molecules existing in other compartments through the ultrathin walls: Resultant uniqueness may firstly be an elevated freezing temperature brought by the strongly overlapping pore-wall potential, and secondly a possibility of the cooperative phase transitions even in sub-nano pore space, which may not be the case for fluids in usual micropores. Another uniqueness may result from packing effect of molecules: the phase behavior would be extremely sensitive to the Å-order of difference in the channel size. Preceding works of molecular simulations for MOF pore systems of course exist [7-9], but the above kind of viewpoints seem lacking. This work aims at, NOT mimicking or expressing adsorption isotherms, BUT finding basic feature of fluids confined in this new type of porous materials.

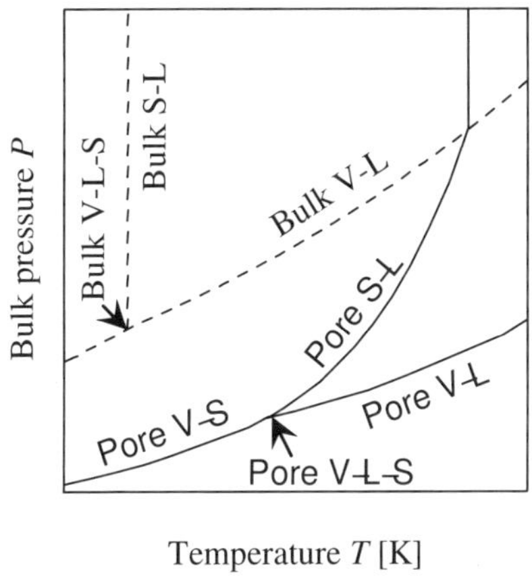

Figure 1. Whole phase diagram of LJ fluid confined in slit nanospace with strongly attractive walls.

Until now, not many results have yet been obtained, but we have made some GCMC simulations for LJ-methane fluid confined in quasi-1D channels and in the jungle-gym space. What have been found for the former case are: i) extremely elevated freezing points for quasi-1D channels made up of graphene sheets as the walls, and ii) enhancement of freezing even with walls made up of fluid molecules themselves, which cannot be the case for SLIT geometry with walls made up of fluid molecules. As for the latter nanospace, hindrance of freezing and acceleration of condensation by the jungle-gym structure are observed until now, which will not be shown in this paper but be discussed in the conference.

2. GCMC Simulation

The GCMC method was employed, with which the bulk-phase state in equilibrium in the pore system can be clarified. The potential model for fluid-fluid interaction was Lennard-Jones (LJ) 12-6 function modeled for methane (ε_{ff}/k = 148.1 K, σ_{ff} = 0.381 nm). The cut-off distance was $5\sigma_{ff}$, which was thought to be large enough to represent fluid with the full LJ potential. Thus no long-range correction was attempted.

The unit cell was composed of N times N array of quasi-1D channels with given size, each of which was divided by single atomic layer represented by LJ 10-4 potential function

$$\phi_{fs}^{10-4}(z) = 2\pi\varepsilon_{fs}\sigma_{fs}{}^2 N_c \left[\frac{2}{5}\left(\frac{\sigma_{fs}}{z}\right)^{10} - \left(\frac{\sigma_{fs}}{z}\right)^4\right]. \quad (1)$$

A fluid molecule in a channel receives not only the above fluid-solid interaction but also those from fluid molecules within the cutoff distance, some of which may exist in other compartments beyond the ultrathin walls. This is the reason the unit cell contains N-by-N array of channels. The number N was at least three or more, determined so as to satisfy the usual condition of (Unit cell length)/2 > (Cutoff distance). Though the molecules themselves never go beyond the unit cell along the confining direction, the periodic boundary conditions and the minimum image convention for all the three directions were set in the simulations to take the above explained interactions into account.

The LJ parameters for solid employed was i) those for graphene sheet and ii) those for *methane* sheet that corresponds to single (111) layer of fcc solid methane at triple point. The latter is useful for extracting the geometrical effect of the pore system [2]. The Lorentz-Berthelot mixing rules were used to evaluate solid-fluid interaction parameters.

A correction of fluid-solid interaction must be made about the intersection of the lateral and vertical walls: Simple sum of the two would overestimate twice the real interaction from the overlapping portion. The correction was possible by subtracting a LJ 11-5 potential:

$$\phi_{fs}^{11-5}(r) = \frac{3\pi\varepsilon_{fs}\sigma_{fs}N_l}{2}\left[\frac{21}{32}\left(\frac{\sigma_{fs}}{r}\right)^{11} - \left(\frac{\sigma_{fs}}{r}\right)^{5}\right], \tag{2}$$

which was derived from line-integration of LJ 12-6 potential.

The system traced the gas-liquid coexistence line for bulk fluid (and gas-solid one if depression was the case), which corresponds to the pore system immersed in liquid or solid. The coexistence T-μ relations [2] were used as inputs to the simulations. A few to several hundred millions of elemental GCMC steps (movement, insertion or deletion) were conducted for each condition.

3. Results and Discussion

Some examples of simulation results are shown in Figures 2 and 3. The LJ-methane in the graphitic walls with the effective channel size of ca. 4σ exhibits solid-like structure even at as high a temperature as 185 K, or near the bulk critical temperature, which is demonstrated by the hexagonal arrangement of the molecules in the layer contacting to the walls (Figure 2) and by the almost flat plateau in density upon further cooling (Figure 3).

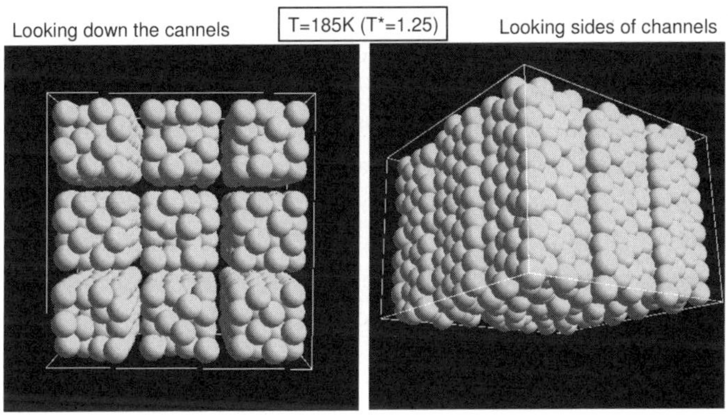

Figure 2. Quasi-solid phase observed in coordination space even around the bulk critical temperature

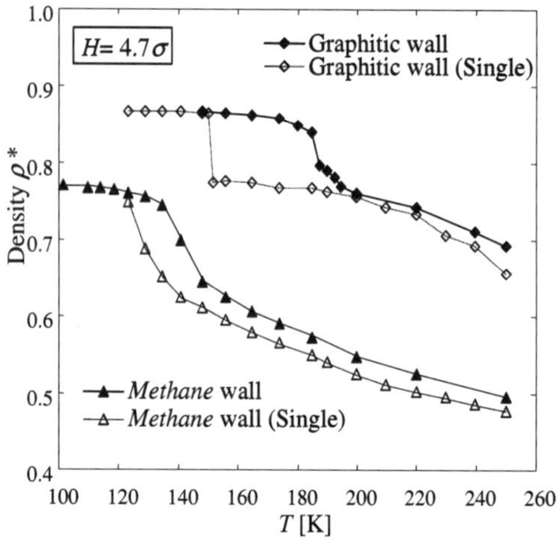

Figure 3. Density variation in the channels

We have tried to characterize the structure by any statistic information, and found that the pair correlation function can be extracted layer-by-layer, even for this kind of strongly anisotropic structure of molecules. Figure 4 shows the in-plane pair correlation function for the contacting layer, which demonstrates that the structure at the higher temperature is rather liquid-like with random nature. On the other hand a decrease in temperature down to 185 K brings almost perfect hexagonal order, which is typically demonstrated by the first minimum reaching down to zero, and by the sprit of the second peak.

For the isolated channel (noted as "Single") in Figure 3, the freezing occurs at a lower temperature than the bundle of the channels, which is clear indication of the importance of fluid-fluid interaction across the thin walls.

Possible origin of the solidification at such a high temperature would firstly be the superposition of potential energies from surrounding solid walls. However, this factor alone cannot explain the results for *Methane*-wall case, in which freezing occurs at a higher temperature than the bulk freezing point for LJ-methane (ca. 100K): Since the walls have only the same interaction strength as those between fluid molecules, simple superposition of such potential alone would not accelerate solidification, which had been demonstrated in the case for slit geometry [2]. Thus another factor seems to be existing and enhancing the freezing in the channels with the size of a few times the molecular diameter. The most likely candidate would be reduction in mobility or suppression of local density fluctuation brought by strong and narrow confinement in two of the space directions. Some arbitrariness, however, may stand in the choice of crystal face for the *Methane*-wall, and additional examination would be necessary before ensuring the existence of the above factor.

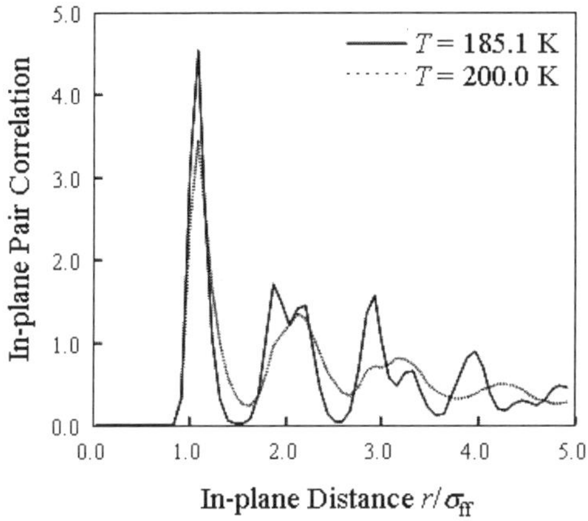

Figure 4. Pair correlation functions for 185K (solid-like) and 200K (liquid-like).

Further study on effects of the size of channel and interaction strength of wall is expected to give detailed understanding of the phase behavior in MOF spaces. Also highly desired is development of the study for examining cooperativeness of framework transition, or the gate effect, in near future.

4. Conclusion

Towards the exploration and understanding for phase behavior of simple fluids confined in coordination nanospace, or so-called the metal-organic framework, freezing behavior of LJ-methane in array of quasi-1D channels was examined employing GCMC technique. A unit cell that contains at least 3x3 array of square channels divided by thin walls of single atom thickness was developed and the followings have been clarified through the simulations. i) The LJ-methane in the graphitic walls with the effective cannel size of ca. 4σ exhibited solid-like structure even at as high a temperature as around the bulk critical temperature of 185 K, ii) Comparison with the isolated single channel demonstrated significance of the fluid-fluid interaction beyond the thin walls, iii) Unlike the case with slit geometry, the walls made up with fluid molecules themselves still exhibited elevated freezing point than that for bulk fluid, which implied existence of enhancing factor for freezing that was not prevailing slit-pore case.

Acknowledgements

This work was supported in part by the Grant-in-Aid for Scientific Research on Priority Areas, "Chemistry of Coordination Space", by MEXT, Japan.

References

1. Miyahara M., Kanda H., Yoshioka T. and Okazaki M., Modeling capillary condensation in cylindrical nanopores: a molecular dynamics study, *Langmuir* **16** (2000) pp. 4293–4299.
2. Miyahara M. and Gubbins K. E., Freezing/melting phenomena for Lennard-Jones methane in slit pores: a Monte Carlo study, *J. Chem. Phys.* **106** (1997) pp. 2865–2880.
3. Kanda H., Miyahara M. and Higashitani K., Solidification of Lennard-Jones fluid in cylindrical nanopores and its geometrical hindrance effect: a Monte Carlo study, *Langmuir*, **16** (2000) pp. 8529–8535.
4. Miyahara M., Kanda H., Shibao M. and Higashitani K., Solid-liquid phase transition of Lennard-Jones fluid in slit pores under tensile condition, *J. Chem. Phys.* **112** (2000) pp. 9909–9916.
5. Kanda H., Miyahara M. and Higashitani K., Triple point of Lennard-Jones fluid in slit pore – solidification of critical condensate –, *J. Chem. Phys.* **120** (2004) pp. 6173–6179.
6. Kanda H., Miyahara M. and Higashitani K., Sublimation phenomena in slit nanopores: Lennard-Jones phase diagram, *Adsorption* **11** (2005) pp. 295–299.
7. Bojan M. J. and Steele W. A., Computer simulation in pores with rectangular cross-sections, *Carbon* **36** (1998) pp. 1417–1423.
8. Vishnyakov A., Ravikovich P. I., Neimark A.V. Bulow M. and Wang Q. M., Nanopore structure and sorption properties of Cu-BTC metal-organic framework, *Nano Let.* **3** (2003) pp. 713–718.
9. Duren T., Sarkisov L., Yaghi O. M. and Snurr R. Q., Design of new materials for methane storage, *Langmuir* **20** (2004) pp. 2683–2689.

EQUILIBRIUM THEORY-BASED DESIGN OF SMBS FOR A GENERALIZED LANGMUIR ISOTHERM

MARCO MAZZOTTI

ETH Zurich, Institute of Process Engineering, Sonneggstrasse 3, CH-8092 Zurich, Switzerland

E-mail: mazzotti@ipe.mavt.ethz.ch

This work presents design criteria for complete separation of binary mixtures in Simulated Moving Bed (SMB) separations that apply to systems, whose retention behavior is characterized by a generalized Langmuir isotherm. By allowing for negative terms in the denominator of the classical Langmuir isotherm, this newly introduced adsorption model covers a broad class of adsorption isotherms, including Langmuir or anti-Langmuir behavior for both adsorbates, and mixed cases where one species behaves in a Lagmuirian and the other in an anti-Langmuirian manner. By extending classical equilibrium theory results for the binary Langmuir isotherm, and by generalizing the approach followed earlier to derive SMB design criteria for the binary and multi-component Langmuir isotherm, exact algebraic equations for the boundary of the complete separation region in the operating parameter space are derived for all possible generalized Langmuir isotherm.

1. Introduction

Simulated Moving Beds (SMBs) are well established for the adsorption based separation of hydrocarbons as well as of fine chemicals, particularly enantiomers. This technology covers a broad range of production scales from the laboratory units, which use chromatographic columns with a 0.5 cm internal diameter, to the multi-ton production units licensed by Novasep for chiral separations with column diameters between 20 and 100 cm, to the largest SMB unit licensed recently in South Korea by the Institute Francaise du Petrol with a column diameter of 8 m for the production of 700,000 tons per year of p-xylene. New applications are envisaged in the near future, particularly in the emerging area of bio-separations, e.g. for the purification of enzymes, peptides, antibiotics and natural extracts.

The design of SMB units for such a wide range of applications requires the use of models of different levels of complexity. Detailed models are typically used for simulation and optimization, whereas Equilibrium Theory based models are used for design purposes, yielding the so-called Triangle Theory that was developed and is used for systems whose adsorption is characterized by the Langmuir isotherm [1,2].

In this work we present an extension of the Equilibrium Theory and of the Triangle Theory to a more general class of isotherms that we call, generalized Langmuir isotherm.

2. Generalized Langmuir isotherm

The binary systems considered in this work are characterized by a generalized form of the Langmuir isotherm, which is defined as follows [3]:

$$n_i = \frac{H_i c_i}{1 + p_A K_A c_A + p_B K_B c_B} = \frac{H_i c_i}{\delta} \qquad (i = A, B)$$

where c_i and n_i are fluid and adsorbed phase concentrations, respectively; K_i and H_i are the equilibrium constant and the Henry's constant, respectively ($H_A > H_B$, i.e. the second component is more retained than the first). Note that the denominator δ must be positive for the isotherm to have physical meaning. The parameters p_A and p_B can take the values ± 1 and characterize the Langmuir or anti-Langmuir character of the behavior of the corresponding species. The Langmuir isotherm (indicated as case **L** in the following) is obtained in fact when $p_A = p_B = 1$. A synergistic anti-Langmuir isotherm, case **A**, is obtained when $p_A = p_B = -1$. Two mixed isotherms combinations are also possible, namely the mixed case $\mathbf{M_1}$ where $p_B = 1 = -p_A$, and the mixed case $\mathbf{M_2}$ where $p_A = 1 = -p_B$.

The latter mixed isotherm, case $\mathbf{M_2}$, is special because the mathematical model equations are mixed hyperbolic-elliptic partial differential equations [3], and the analysis presented here is valid only in the region of the composition space close to the origin where the equations are hyperbolic, i.e. when the following additional constraints are fulfilled:

$$c_B^F < k; \quad c_A^F < k/h; \quad 0 < \left(c_B^F - k + hc_A^F\right)^2 - 4hc_A^F c_B^F;$$
where $h = (K_A H_B)/(K_B H_A)$ and $k = (1 - H_B/H_A)/K_B$.

3. Equilibrium Theory for the generalized Langmuir isotherm

The Equilibrium Theory of chromatography is a very powerful tool to study and understand the dynamics of chromatographic columns for single component, binary and multi-component systems, whose retention behavior is described by any type of isotherm. The mathematical model equations are solved using the method of characteristics, and in the case of the Langmuir isotherm one finds out that the characteristics are straight lines in the composition space, thus allowing for a quite simple closed-form solution in many cases of practical interest [1]. We have recently extended these classical results to binary systems described by the generalized Langmuir isotherm reported above. We have demonstrated that in all four cases the characteristics are straight lines in the composition space, which are the tangents to a parabola [3]. Moreover, Riemann problems, i.e. piecewise constant initial value problems, have solutions that can be obtained using concepts and methods similar to those used for the Langmuir isotherm. As illustrated in Figure 1, the parabola for each of the four cases belongs to a different quadrant in the (c_A,c_B) plane, and the topology of the straight characteristics is accordingly different; all the details have been reported elsewhere [3]. It is worth noting the striking symmetry of the characteristic fields in the composition space for the different generalized Langmuir siotherms.

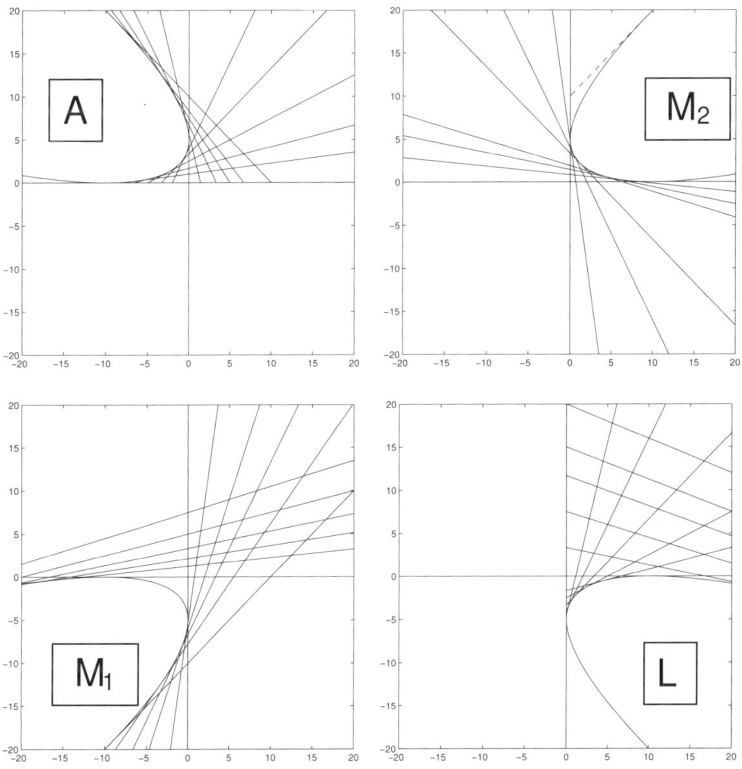

Figure 1. Characteristic fields in the (c_A,c_B) plane (c_A is the horizontal coordinate).

In the frame of the Equilibrium Theory an important role is played by the one-to-one mapping between the composition space, i.e. the (c_A,c_B) plane, and the space of the characteristic parameters, i.e. the (ω_1,ω_2) plane. With reference to the composition of the feed stream in a SMB unit for instance, the corresponding pair of ω values is obtained by solving the following quadratic equation:

$$\left(1+p_A K_A c_A^F + p_B K_B c_B^F\right)\omega^2 - \left[H_B\left(1+p_A K_A c_A^F\right)+H_A\left(1+p_B K_B c_B^F\right)\right]\omega + H_A H_B = 0$$

It can be demonstrated that the ω values fulfill the following inequalities [3]:

$$\text{case L}: \quad 0<\omega_1^F \le H_B \le \omega_2^F \le H_A$$

$$\text{case A}: \quad H_B<\omega_1^F \le H_A \le \omega_2^F < \infty$$

$$\text{case M}_1: \quad 0<\omega_1^F \le H_B < H_A \le \omega_2^F < \infty$$

$$\text{case M}_2: \quad H_B \le \omega_1^F \le \omega_2^F \le H_A$$

4. Triangle Theory for the generalized Langmuir isotherm

In this work we consider a four-section Simulated Moving Bed (SMB) unit, where a binary mixture is separated in such a way to achieve complete separation, i.e. to collect only component 1 pure in the Raffinate, and only component 2 pure in the Extract. In the frame of Equilibrium Theory SMB separation performances depend on the dimensionless flow rate ratios m_j that are defined as follows in terms of the operating parameters of the SMB:

$$m_j = \frac{Q_j t^* - \varepsilon^* V}{V(1-\varepsilon^*)} \qquad (j=1,\ldots,4)$$

where Q_j is the volumetric flow rate in section j of the SMB; t^* is the switch time, i.e. the time period between two successive switches of the inlet and outlet ports of the SMB; V is the volume of one column in the SMB; ε^* is the overall column void fraction.

The equilibrium theory has been extensively used to design SMB separations, leading to what is sometimes called Triangle Theory; its main application has been so far to systems characterized by the Langmuir isotherm [2,4]. Triangle Theory has helped not only to better design but also to better understand SMB separations.

It has recently been possible to extend Triangle Theory to the generalized Langmuir isotherm [5]. Simple algebraic equations that define the region of complete separation in the operating parameter space have been obtained. The mathematical tools and the detailed derivations have been reported elsewhere, and this work provides a compendium of the results to be used even without being familiar to the mathematical techniques behind them.

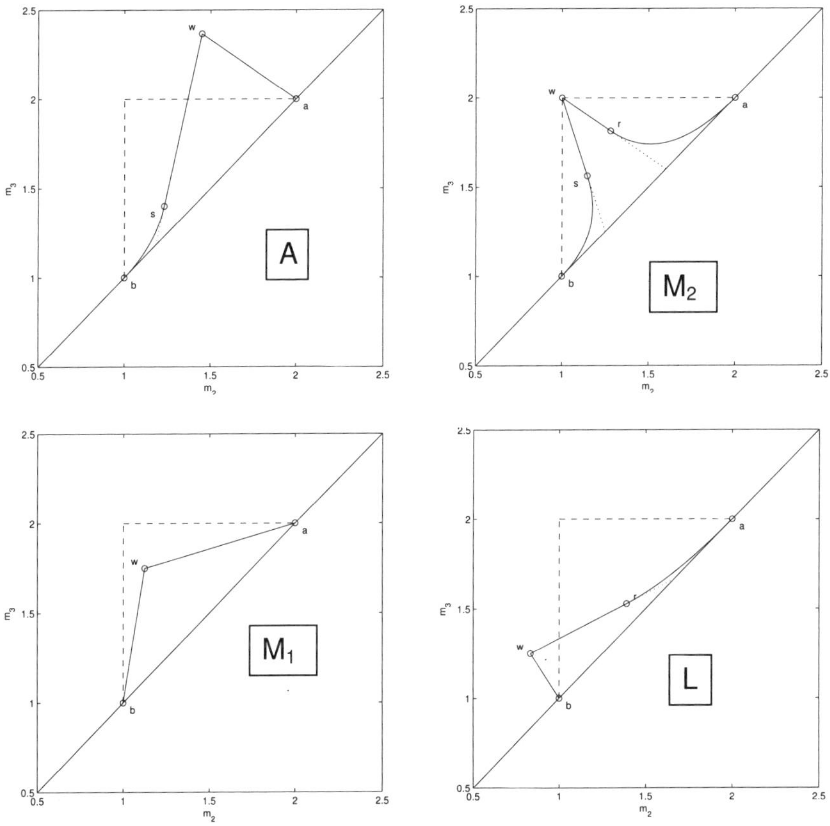

Figure 2. Region of complete separation in the (m_2,m_3) plane. Parameters used are H_A=2, H_B=1, K_A=K_B=0.1 L/g; feed composition: c_A=c_B = 2 (case **A**), 1.5 (case **M**$_2$), 5 (case **M**$_1$), and 4 (case **L**) g/L.

In the case of sections 1 and 4, the constraints on the flow rate ratios to achieve complete separation are explicit and are given by the following relationships:

$$m_1 \geq \frac{1}{2}\left\{m_2 + H_A + K_A c_A^F (m_3 - m_2) + \sqrt{[m_2 + H_A + K_A c_A^F (m_3 - m_2)]^2 - 4m_2 H_A}\right\} \quad (\mathrm{A,M_1})$$

$$m_1 \geq H_A \quad (\mathrm{L,M_2})$$

$$m_4 \leq \frac{1}{2}\left\{m_3 + H_B + K_B c_B^F (m_3 - m_2) - \sqrt{[m_3 + H_B + K_B c_B^F (m_3 - m_2)]^2 - 4m_3 H_B}\right\} \quad (\mathrm{L,M_1})$$

$$m_4 \leq H_B \quad (\mathrm{A,M_2})$$

Note that different inequalities apply to different isotherms as indicated and that in two of these the bounds depend on the flow rate ratios in sections 2 and 3, on the adsorption isotherm parameters and on the feed composition.

Table 1. Intersection points on the boundary of the complete separation regions in Figure 2. Note that in this table the subscripts 1 and 2 replace subscripts *B* and *A*, respectively, that have been used in all other equations.

Point	m_2	m_3
a	H_2	H_2
b	H_1	H_1
f	ω_2^F	ω_2^F
g	ω_1^F	ω_1^F
r	$\frac{(\omega_2^F)^2}{H_2}$	$\frac{\omega_2^F[\omega_2^F(H_1-\omega_1^F)+\omega_1^F(H_2-H_1)]}{H_1(H_2-\omega_1^F)}$
s	$\frac{\omega_1^F[\omega_1^F(\omega_2^F-H_2)+\omega_2^F(H_2-H_1)]}{H_2(\omega_2^F-H_1)}$	$\frac{(\omega_1^F)^2}{H_1}$
$\mathbf{w}_0$ (linear case)	H_1	H_2
$\mathbf{w}_L$ (case **L**)	$\frac{\omega_2^F H_1}{H_2}$	$\frac{\omega_2^F[H_1(H_1-\omega_1^F)+\omega_1^F(H_2-H_1)]}{H_1(H_2-\omega_1^F)}$
$\mathbf{w}_A$ (case **A**)	$\frac{\omega_1^F[H_2(\omega_2^F-H_2)+\omega_2^F(H_2-H_1)]}{H_2(\omega_2^F-H_1)}$	$\frac{\omega_1^F H_2}{H_1}$
$\mathbf{w}_{M_1}$ (case $\mathbf{M}_1$)	$H_1\left\{1+\frac{(H_2-\omega_1^F)(\omega_2^F-H_2)(H_2-H_1)}{H_2[(H_1-\omega_1^F)(\omega_2^F-H_2)+(H_2-\omega_1^F)(\omega_2^F-H_1)]}\right\}$	$H_2\left\{1-\frac{(H_1-\omega_1^F)(\omega_2^F-H_1)(H_2-H_1)}{H_1[(H_1-\omega_1^F)(\omega_2^F-H_2)+(H_2-\omega_1^F)(\omega_2^F-H_1)]}\right\}$
$\mathbf{w}_{M_2}$ (case $\mathbf{M}_2$)	$\frac{\omega_1^F\omega_2^F}{H_2}=\frac{H_1}{\delta_F}$	$\frac{\omega_1^F\omega_2^F}{H_1}=\frac{H_2}{\delta_F}$

In the case of sections 2 and 3, the constraints on the flow rate ratios are coupled and define a two-dimensional region in the (m_2,m_3) plane. The complete separation regions for the four cases of generalized Langmuir isotherm are shown in Figure 2, together with the region for the linear isotherm with the same Henry's constants as the generalized Langmuir isotherm.

The equations for the straight lines can be derived by the coordinates of the intersection points that are reported in Table 1. The equations of the only two curves on the boundaries of the complete separation regions are as follows:

$$m_3 = m_2 + \left(\sqrt{m_2} - \sqrt{H_A}\right)^2 \Big/ \left(p_A K_A c_A^F\right) \qquad \text{(line ar)}$$

$$m_2 = m_3 + \left(\sqrt{m_3} - \sqrt{H_B}\right)^2 \Big/ \left(p_B K_B c_B^F\right) \qquad \text{(line bs)}$$

Also in Figure 2, as in Figure 1, a remarkable and a striking symmetry among the four different cases can be recognized.

5. Conclusions

In this paper recent results about the design of SMB separations for a new type of isotherm, i.e. the generalized Langmuir isotherm, have been summarized. This represents a significant advancement is the field of SMB modelling, design and optimization, and it is expected to have an impact also on applications. The results that have been obtained through Equilibrium Theory are cast in a simple form that makes their use rather straightforward. They allow for a deep understanding of SMB operation for non-Langmuir binary isotherms, particularly for a clarification of the effect of operating parameters and of feed composition on the shape and position of the complete separation region in the (m_2,m_3) plane [6].

References

1. Rhee H-K., Aris R., Amundson N. R., *First order partial differential equations, vol 2,* Prentice-Hall, Englewood Cliffs, New Jersey (1989).
2. Storti G., Mazzotti M., Morbidelli M., Carrà S., Robust design of binary countercurrent adsorption separation processes, *AIChE J.* **39** (1993) pp. 471-492.
3. Mazzotti M., Local equilibrium theory for the binary chromatography of species subjected to a generalized Langmuir isotherm, *Ind. Eng. Chem. Res.* **45** (2006a) pp. 5232-5350.
4. Chiang A.S.T., Complete separation conditions for a local equilibrium TCC adsorption unit, *AIChE J.* **44** (1998) pp. 332-340.

5. Mazzotti M., Design of Simulated Moving Bed separations – Generalized Langmuir isotherm, *Ind. Eng. Chem. Res.* **45** (2006b) pp. 6311-6324.
6. Mazzotti M., Equilibrium theory based design of Simulated Moving Bed processes for a generalized Langmuir isotherm, *J. Chrom. A* **1126** (2006c) pp. 311-322.

NON-EQUILIBRIUM DYNAMIC ADSORPTION AND DESORPTION ISOTHERMS OF CO_2 ON A K-PROMOTED HTLC

STEVEN P. REYNOLDS, ARMIN D. EBNER AND JAMES A. RITTER

Department of Chemical Engineering, University of South Carolina, Columbia, SC 29208, USA

E-mail: ritter@engr.sc.edu

A K-promoted HTlc was synthesized and tested for its reversible CO_2 capacity between 250 and 500 °C. Non-equilibrium dynamic adsorption and desorption isotherms were measured between 65 and 980 torr using 20 or 50 torr steps and a 45 min duration between steps. The absolute CO_2 capacity on K-promoted HTlc increased with decreasing temperature, with CO_2 loadings of 2.25 and 1.02 mol/kg respectively at 250 and 500 °C and 980 torr. The reversible CO_2 working capacity obtained between 65 and 980 torr exhibited a maximum at 450 °C, with a value of 0.55 mol/kg compared to 0.11 and 0.46 mol/kg at 250 and 500 °C, respectively. It was surmised that three temperature dependent, highly coupled, completely reversible, equilibrium driven but kinetically limited reactions were taking place, with the first one being a rapid and reversible chemisorption of CO_2 that initiated the entire process.

1. Introduction

The economic capture and concentration of CO_2 from flue gas is a daunting challenge [1]. Chemical and physical absorption, cryogenic distillation, membrane, and chemical and physical adsorption processes are all being investigated and developed for this purpose [1]. However, a cost effective CO_2 separation technology has not been identified [1].

Various adsorption processes have been proposed for CO_2 capture and concentration [1]. One of the more promising approaches considers the use of a pressure swing adsorption (PSA) process at high temperature [2,3]. This PSA process is based on the use of a K-promoted hydrotalcite like compound (HTlc) that exhibits a reversible capacity for CO_2 at elevated temperatures [4]. However, a paucity of information is available on HTlc materials, especially for reversible CO_2 adsorption [5-8].

The objective of this article is to report on a K-promoted HTlc that is being touted as a high temperature CO_2 adsorbent [4]. This material was synthesized [4] and then studied to determine its reversible CO_2 capacity at elevated temperatures. Because this material took excessive time to equilibrate, but exhibited complete reversibility with CO_2 [9], the results from non-equilibrium dynamic cycling experiments are reported that elucidate the adsorption and desorption behavior of CO_2 on K-promoted HTlc when exposed to various temperatures and CO_2 pressures for finite periods of time.

2. Adsorbent Preparation and Isotherm Measurement

A HTlc with molecular formula $[Mg_3Al(OH)_8]_2CO_3 \bullet nH_2O$ was prepared by a co-precipitation method [4]. While vigorously stirring, a solution of 41.7 ml of deionized water containing 0.75 mol $Mg(NO_3)_2 \bullet 6H_2O$ and 0.25 mol $Al(NO_3)_3 \bullet 9H_2O$ was added to a solution of 83.3 ml of deionized water containing 1.7 mol NaOH and 0.5 mol Na_2CO_3. The precipitate was separated from the slurry by vacuum filtration. The wet filter cake was washed with deionized water and vacuum filtered three times, dried overnight at 60 °C in a vacuum oven, crushed, and calcined in air at 400 °C for 4 hours.

A K-promoted HTlc was prepared using an incipient wetness procedure. To obtain a Al:K ratio of 1:1, a 0.33 M solution of K_2CO_3 was prepared in deionized water, and a pre-determined volume of it was added to the HTlc powder in three steps: 1) The solution was added drop wise to the powder until it appeared wet. 2) The wet powder was dried for 15 min in a vacuum oven at 60 °C. 3) Steps 1 and 2 were repeated until all the solution was added.

A VTI Integrated Microbalance system was utilized to measure the non-equilibrium dynamic adsorption and desorption isotherms of CO_2 on the K-promoted HTlc. For each isotherm, ~ 0.1 g of sample was loaded into the microbalance, evacuated to 1×10^{-5} torr, and activated in vacuum at 400 °C for 12 hours. After activation, the temperature was changed to the isotherm temperature ($\pm$ 1 °C) for subsequent contact with CO_2.

A non-equilibrium adsorption and desorption isotherm at 250, 300, 350, 400, 450 or 500 °C was measured by taking differential pressure steps of 20 $\pm$ 5 torr between 65 and 300 torr and 50 $\pm$ 5 torr between 300 and 980 torr (27 steps up and 27 steps down), waiting 45 min at each step, and proceeding in this manner until periodic behavior was realized. This produced Langmuirian-shaped isotherms under non-equilibrium conditions. The absolute and the dynamic working capacities of CO_2 on K-promoted HTlc were extracted from these non-equilibrium isotherms.

3. Results and Discussion

Figure 1 shows the non-equilibrium dynamic adsorption and desorption isotherms at all six temperatures for CO_2 on K-promoted HTlc at the periodic state. Depending on the temperature, between 5 and 12 adsorption and desorption cycles were required in each case to attain periodic behavior [9]. The approach to periodic behavior was associated with an initial non-equilibrium CO_2 capacity that exhibited substantial departure not only from equilibrium but also from the periodic absolute CO_2 adsorption capacity, with this departure being larger with decreasing temperature [9]. A hysteresis loop formed between the non-equilibrium dynamic adsorption and desorption isotherms and remained intact at the periodic state.

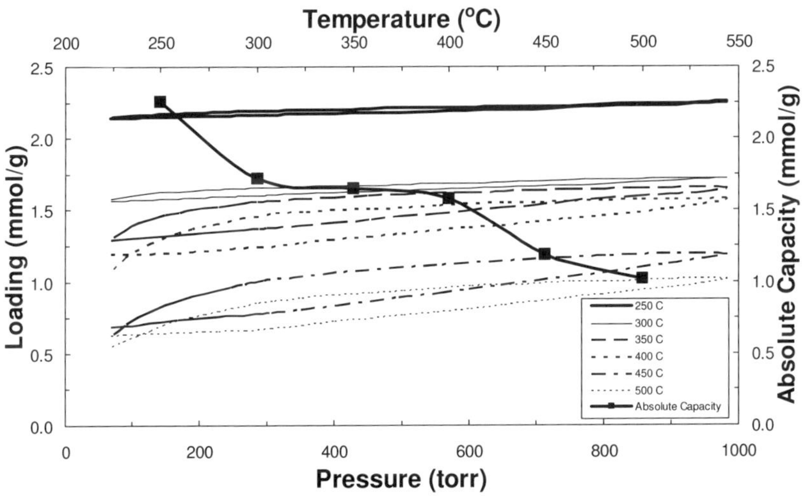

Figure 1. Dynamic non-equilibrium adsorption and desorption isotherms at 250, 300, 350, 400, 450 and 500 °C for CO_2 on K-promoted HTlc at the periodic state; and non-equilibrium absolute capacity for CO_2 on K-promoted HTlc obtained from these results at 980 torr.

The corresponding temperature dependence of the absolute CO_2 capacities on K-promoted HTlc obtained from these results at 980 torr is also shown in Figure 1. This capacity initially decreased with increasing temperature, reached a plateau at around 300 to 400 °C, and then decreased again with further increases in the temperature. This behavior was indicative of an exothermic adsorption mechanism because of the increasing CO_2 capacity with decreasing temperature;

and the plateau was perhaps caused by a phase transition occurring within the material that approached a critical temperature at around 500 °C. This absolute CO_2 capacity ranged from 1.02 mol/kg at 500 °C to 2.25 mol/kg at 250 °C. These CO_2 capacities and temperature trends were comparable with those reported elsewhere [4-8].

The results from Figure 1 are re-plotted in Figure 2 in terms of the CO_2 loading normalized to 0.0 mol/kg at 65 torr. It was now easy to observe not only the significant changes in the CO_2 loadings, but also the marked changes in the sizes of the hysteresis loops, that occurred between 65 and 980 torr with temperature. The temperature dependence of the CO_2 working capacity, defined here as the CO_2 loading change between 65 and 980 torr of each isotherm is also shown in Figure 2. The CO_2 working capacity exhibited strong temperature dependence and a maximum of 0.55 mol/kg at around 450 °C. Below this temperature it decreased almost linearly down to 0.11 mol/kg at 250 °C, and above this temperature it also decreased down to 0.46 mol/kg at 500 °C. The larger hysteresis loops with increasing CO_2 working capacity were counterintuitive but consistent with faster desorption kinetics in the low pressure regions being offset by relatively slower desorption kinetics in the high pressure regions. These results perhaps indicated that two fundamentally different phenomena associated with two different interchangeable CO_2 phases were taking place within the K-promoted HTlc structure.

Based on the culmination of these findings, the following mechanism was envisioned for the reversible uptake and release of CO_2 in K-promoted HTlc. The decreasing absolute CO_2 capacity with increasing temperature was consistent with an equilibrium driven, exothermic process (reaction). This absolute CO_2 capacity was most likely associated with a high capacity, reversible, CO_2 phase (phase C) that exhibited relatively slow adsorption and desorption (or reaction) kinetics.

The CO_2 working capacity that generally increased with increasing temperature, but that exhibited a maximum at high temperatures, was probably associated with a different CO_2 phase (phase B). This phase exhibited an intermediate and reversible CO_2 capacity and relatively fast adsorption and desorption (reaction) kinetics. It was also deduced that the reason Phase B exhibited an increase in capacity with increasing temperature (i.e., the CO_2 working capacity) was due to phase C losing capacity that was made available to phase B. The fact that phase B eventually lost capacity with increasing temperature after exhibiting a maximum suggested that it was also associated with an exothermic process.

It was further envisioned that phases B and C were coupled to each other through an equilibrium driven, but kinetically limited, reversible reaction that was very sensitive to temperature. Also, phase B was formed from the reversible conversion of a weakly bound chemisorbed layer of CO_2 (phase A). This phase was responsible for the rapid adsorption and desorption kinetics in the low pressure regions and was not as sensitive to temperature [9].

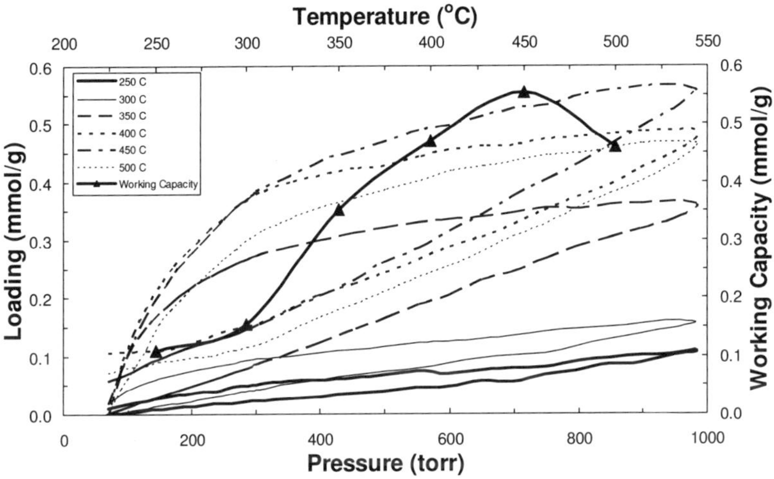

Figure 2. Non-equilibrium dynamic adsorption and desorption isotherms at 250, 300, 350, 400, 450 and 500 °C for CO_2 on K-promoted HTlc at the periodic state, with each isotherm from Figure 1 normalized to zero CO_2 loading at 65 torr; and non-equilibrium dynamic working capacities for CO_2 on K-promoted HTlc obtained from these results between 65 and 980 torr.

4. Conclusions

A K-promoted HTlc was synthesized and tested to determine its reversible CO_2 capacity between 250 and 500 °C. Non-equilibrium dynamic adsorption and desorption isotherms were measured between 65 and 980 torr using 20 and 50 torr steps and a 45 min duration between steps. The absolute CO_2 capacity on K-promoted HTlc increased with decreasing temperature, with CO_2 loadings of 2.25 and 1.02 mol/kg respectively at 250 and 500 °C and 980 torr. The CO_2 working capacity obtained between 65 and 980 torr exhibited a maximum at 450 °C, with a value of 0.55 mol/kg compared to 0.11 and 0.46 mol/kg at 250 and 500 °C, respectively.

Each isotherm exhibited the following characteristics: Depending on the temperature, it took between 5 and 12 adsorption and desorption cycles to attain periodic behavior. The approach to periodic behavior was associated with an initial non-equilibrium CO_2 capacity that exhibited substantial departure not only from equilibrium but also from the periodic absolute CO_2 adsorption capacity, with this departure being larger with decreasing temperature. A hysteresis loop formed between the non-equilibrium dynamic adsorption and desorption isotherms and remained intact at the periodic state.

These results were interpreted in terms of the uptake and release of CO_2 on K-promoted HTlc being associated with three temperature dependent, coupled, reversible and equilibrium driven reactions. The third reaction exhibited slow adsorption and desorption kinetics and a very high CO_2 capacity. The second reaction exhibited faster adsorption and desorption kinetics and an intermediate CO_2 capacity. The first reaction exhibited very rapid adsorption and desorption kinetics, with a slightly smaller CO_2 capacity. The first reaction initiated the entire process by forming a chemisorbed layer of CO_2 within the K-promoted HTlc. This layer reversibly converted into a second phase through the second reaction, which reversibly converted into a third phase through the third reaction.

Acknowledgements

The authors gratefully acknowledge financial support provided by DOE through Grant No. DE-FG26-03NT41799.

References

1. Ebner, A. D. and Ritter, J. A., State-of-the-art adsorption and membrane processes for CO_2 production in the chemical and petrochemical industries, *Sep. Sci. Tech.* submitted (2006).
2. Reynolds, S. P., Ebner, A. D. and Ritter, J. A., New pressure swing adsorption cycles for carbon dioxide sequestration, *Adsorption* **11** (2005) pp. 531-536.
3. Reynolds, S. P., Ebner, A. D. and Ritter, J. A., Stripping PSA cycles for CO_2 recovery from flue gas at high temperature using a hydrotalcite-like adsorbent, *Ind. Eng. Chem. Res.* in press (2006).
4. Nataraj, S. et al., "Process for operating equilibrium controlled reactions," Canadian Patent 2,235,928 (1998).
5. Ding, Y. and Alpay, E., Equilibria and kinetics of CO_2 adsorption on hydrotalcite adsorbent, *Chem. Eng. Sci.* **55** (2000) pp. 3461-3474.
6. Ding, Y. and Alpay, E., High temperature recovery of CO_2 from flue gases using hydrotalcite adsorbent, *Trans IChemE* **79** (2001) pp. 45-51.

7. Yong, Z, Mata V. and Rodrigues, A. E. Adsorption of carbon dioxide onto hydrotalcite-like compounds (HTlcs) at high temperature, *Ind. Eng. Chem. Res.* **40** (2001) pg. 204-209.
8. Yong, Z. and Rodrigues, A. E. Hydrotalcite-like compounds as adsorbents for carbon dioxide, *Energy Convers. Mgmt.* **43** (2002) pg. 1865-1876.
9. Reynolds, S. P., Ebner, A. D. and Ritter, J. A. Unpublished results, University of South Carolina (2006).

OPTIMISATION OF ADSORPTIVE STORAGE: THERMODYNAMIC ANALYSIS AND SIMULATION

S. K. BHATIA

Division of Chemical Engineering
The University of Queensland, Brisbane, QLD 4072, Australia
E-mail: sureshb@cheque.uq.edu

ALAN L. MYERS

Department of Chemical and Biomolecular Engineering
University of Pennsylvania, Philadelphia, PA 19104, U.S.A.
E-mail: amyers@seas.upenn.edu

The storage of gases in porous adsorbents is examined here thermodynamically from a systems viewpoint, to derive concrete objective criteria to guide the search for the 'Holy Grail' adsorbent, for which the adsorptive delivery is maximized. It is shown that for ambient temperature storage of hydrogen and delivery between 30 bar and 1.5 bar pressure, for the optimum adsorbent the adsorption enthalpy change is 15.1 kJ/mole, while for methane it is 18.8 kJ/mole. For carbons, an optimum operating temperature of about 115 K is predicted for hydrogen storage, while for methane the optimum temperature for carbons is 254 K. It is also demonstrated that for maximum delivery of the gas the optimum adsorbent must be homogeneous. These results are confirmed with the help of experimental data from the literature, as well as extensive Monte Carlo simulations conducted here using slit pore models of activated carbons and atomistic models of carbon nanotubes.

1. Introduction

One of the key challenges facing the utilisation of alternate fuels is the development of a viable means of storage, particularly in the mobile energy use sector. Compressed gas is a major alternative fuel source, but requires unacceptably high storage pressures while liquefaction requires prohibitively low temperature (e.g. 20 K for hydrogen). For hydrogen, the U.S. Department of Energy (DOE) has set a target of 6 wt% storage to be achieved by 2010 and 9 wt% by 2015, to match the energy density of hydrocarbons. Hydrides are already able to meet these targets [1-3]; however, the high temperature needed for desorption remains a key concern [2,4]. Storage of both hydrogen and methane in clathrate hydrates [5] requires prohibitively high pressures (>120 bar). Consequently, much effort has been devoted to adsorptive storage.

Key to the success of adsorptive storage is the choice of adsorbent and operating condition. Ambient temperature storage has been the goal, but for H_2 less than 1% by weight storage has been attained at this temperature, with numerous adsorbents such as activated carbon granules and fibres [6,7], carbon nanotubes [8,9], and zeolites [10,11] as well as metal organic frameworks [12,13] investigated.

While progress is being made, and capacities gradually improved, albeit still far from target in the case of hydrogen, the drive to meet DOE goals would appear to lack a well-defined objective. Thus, the necessary properties of the 'Holy Grail' adsorbent have not been objectively established. The general (mis)conception is that an adsorbent with a high heat of adsorption is desirable, in order to enhance storage. However, too high an affinity will lead to excessive amount of residual adsorptive on desorption. Thus an analysis of the entire adsorption-desorption cycle is necessary [14]. For carbons the heat of adsorption for hydrogen is typically about 5.8 kJ/mole, while for methane it is about 16 kJ/mole. For other adsorbents the heats are generally smaller. However, it is not known if such values are in the range for which storage cycle operation at ambient temperature is feasible.

Furthermore, is a homogeneous or heterogeneous adsorbent more desirable? Attempts are being made at creating heterogeneities in various ways, such as by alkali metal doping [15],by ball-milling [16], as well as by ion irradiation [17] to enhance adsorption, particularly in carbons, but it is not established if this is an appropriate strategy. Indeed, such defects have largely created chemisorptive trapping sites with desorption temperatures in the range of 600-950 K that are far too high to be of practical interest.

Carbons remain the most attractive candidates for physisorptive storage of both hydrogen and methane, considering their strong adsorption as well as low cost. Here we develop objective criteria for the desired heat of adsorption and level of heterogeneity for optimum performance of the storage delivery cycle. For a given adsorbent the optimum operating temperature of the cycle is also determined based on thermodynamic grounds, and application for the results to slit pore carbons as well SWNT's is discussed, with support from simulation.

2. Thermodynamic Analysis for Optimum Isosteric Heat and Temperature

As discussed above the current search for a suitable adsorbent for storage lacks a well defined objective in terms of the required strength of the adsorption interaction. To this end we consider a homogeneous adsorbent with the Langmuir isotherm, which is suitable at supercritical conditions, especially for weakly interacting gases such as hydrogen. Upon equilibration at storage pressure P_1, the subsequent delivery at exhaustion pressure P_2 is given by

$$D(K, P_1, P_2) = \frac{KP_1 n_m}{1+KP_1} - \frac{KP_2 n_m}{1+KP_2} \tag{1}$$

where K is the equilibrium constant and n_m is the maximum capacity. It is readily determined that, at fixed P_1 and P_2, the delivery, D, is maximum for $K = 1/\sqrt{P_1 P_2}$. Further, $K = e^{\Delta S^o/R} e^{-\Delta H^o/RT} / P_o$, where ΔH^o is the enthalpy change on adsorption, ΔS^o is the entropy change relative to the standard pressure P_o (1 bar), T is temperature and R is the ideal gas constant. It then follows that

$$\Delta H^o{}_{opt} = T\Delta S^o + \frac{RT}{2}\ln\left(\frac{P_1 P_2}{P_o^2}\right) \tag{2}$$

For the adsorption of hydrogen, it may be readily shown that $\Delta S^o \cong -8R$ for a variety of adsorbents [18]. For the delivery cycle reasonable values of adsorption and desorption pressures may be taken as P_1 = 30 bar and P_2 = 1.5 bar respectively, which upon substitution in Eq.(2) yield $\Delta H^o{}_{opt} = -15.1\text{ kJ/mole}$ at T = 298 K. Thus, for optimum delivery of hydrogen between pressures of 30 bar and 1.5 bar at 298 K, an adsorption enthalpy change of -15.1 kJ/mole is desired. The isosteric heat of adsorption of hydrogen on carbons is substantially less, typically about 5.8 kJ/mole. However, if cryogenic conditions are acceptable then one may determine an optimum temperature of operation in the case of activated carbon, for which delivery is maximized. Following Eq.(2), this temperature is obtained as

$$T_{opt} = \frac{\Delta H^o}{[\Delta S^o + (R/2)\ln(P_1 P_2 / P_o^2)]} \tag{3}$$

which provides T_{opt}=114.4 K, for $\Delta H^o = -5.8\text{ kJ/mole}$. Thus, for optimum performance of the delivery cycle using an activated carbon adsorbent an operating temperature of about 115 K is desirable. This is substantially lower than ambient temperature, and demonstrates the futility of current worldwide efforts at improving ambient temperature hydrogen storage capacity of carbons, and other materials with even lower isosteric heat. These conclusions will be further supported with simulations of the delivery in a subsequent section.

The above concepts may also be applied to methane storage. In this case $\Delta S^o \cong -9.5R$ for a variety of adsorbents [18], and Eq. (2) yields $\Delta H^o = -18.82$ kJ/mole for a cycle operating between 30 bar and 1.5 bar at 298 K. This is consistent with values found for methane in carbons, typically about 16 kJ/mole. Consequently, for methane efficient operation of the storage-delivery cycle should be feasible near ambient temperatures. Indeed, Eq. (3) provides an optimal temperature of 253.3 K for carbons.

3. Simulation

To test the above results and determine maximum deliveries from carbons, grand canonical (GCMC) Monte Carlo simulations were performed here for both slit pores and carbon nanotubes, for the case of hydrogen as well as methane storage. The Lennard-Jones model was employed for the fluid-fluid as well as fluid-solid interactions, using the Lorentz-Berthelot mixing rules, and commonly used parameters listed elsewhere [18]. Isosteric heats were estimated in the simulations following the well-known fluctuation formula [18].

For slit pores, the Steele 10-4 potential [19]

$$\phi_{fs}(z,n) = 2\pi\rho_s\sigma_{fs}^2\varepsilon_{fs}\left[\frac{2}{5}\left(\frac{\sigma_{fs}}{z}\right)^{10} - \left(\frac{\sigma_{fs}}{z}\right)^4\right] \tag{4}$$

is used for the interaction with the pore walls, considering single layer walls for maximum surface area. Periodic boundary conditions in the x and y directions were used in the simulations.

Simulations of delivery were also conducted for the case of single walled carbon nanotubes, using an atomistic model of the tube with carbon atoms arranged on the surface of the tube in a hexagonal lattice. Tubes of four different diameters were considered, corresponding to chiral vectors (6,6), (9,6), (9,9) and (10,10), having diameters (measured between centers of carbon atoms) of 0.81 nm, 1.02 nm, 1.22 nm and 1.36 nm respectively. Of these only the (9,6) tube is chiral. The nanotubes were organized on a square lattice, with spacing between tube surfaces of 0.9 nm. The simulations were conducted in a rectangular three dimensional unit cell, with periodic boundary conditions in all three directions.

4. Results and Discussion

Simulations were conducted for hydrogen delivery from slit pore carbons with uniform pore size, between pressures of 30 bar and 1.5 bar. For the calculation, pore densities from simulation, based on center-to-center pore volume, were converted to specific amounts (per unit mass of carbon) using the specific center-to-center pore volume (in cm^3/g) [18]

$$v = 1.315H \tag{5}$$

Figure 1 (a) depicts the results for the absolute delivery from the micropores as a function of temperature for several slit widths. Clear evidence of an optimum temperature for maximum delivery at any slit width is seen, supporting the earlier analysis, with the optimum temperature decreasing with increase in slit width. This is to be expected, because of the decrease in isosteric heat with slit width. Further, at pore widths of 0.9 nm or 1.08 nm, that are typical for activated carbons, the optimal temperature is about 100 K, which is consistent with our earlier determination of 115 K as being optimal for carbons. Figure 1 (b) depicts the variation of isosteric heat with temperature for the different slit widths, and the locus of the optimum, following Eq. (2). Based on our analysis, the intersection of the latter with the isosteric heat curve at any size provides the optimal temperature at that size. This is readily confirmed for the three smaller sizes, by comparison with the temperatures of maximum delivery in Figure 1 (a).

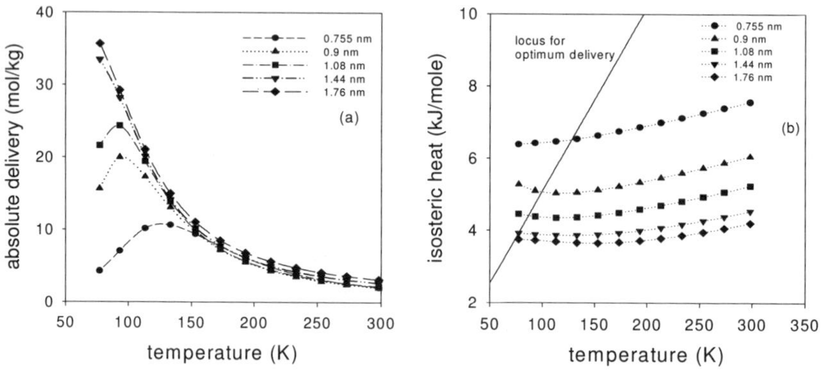

Figure 1. Temperature variation of (a) specific absolute delivery, (b) isosteric heat of adsorption, for hydrogen on activated carbons of various pore sizes.

Figure 2 depicts the results of simulations of hydrogen delivery from carbon nanotubes packed in a square array, and spaced 0.9 nm apart. In all the nanotubes of different sizes examined it is seen that the optimal temperature is significantly reduced, and less than 77 K. This is due to the highly inhomogeneous nature of the interstitial pore space in the nanotube array, which is increasingly filled at the low temperatures.

In comparison to slit pore activated carbons, where higher optimal temperatures have been found, it would appear that carbon nanotubes are less attractive. Indeed, even the absolute deliveries of about 23 mole/kg or 4.6 wt.% at 100 K are lower than the amounts of about 28 mole/kg, or 5.7 wt % obtained for activated carbons at this temperature. Nevertheless, it will be shown subsequently that that the nanotubes in the square array chosen here make more efficient use of the space.

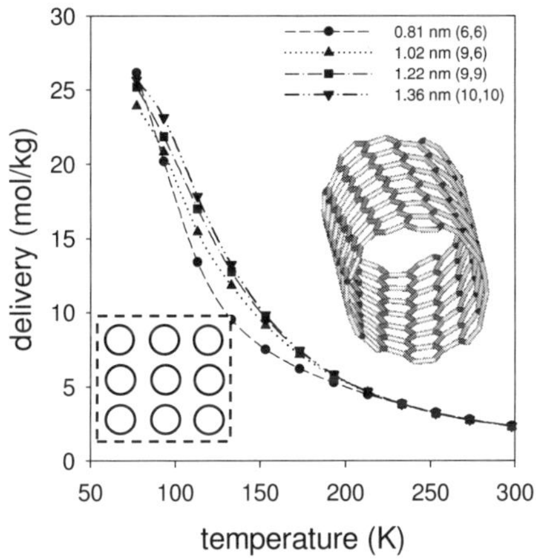

Figure 2. Temperature variation of specific absolute delivery for hydrogen on activated carbons of various pore sizes.

For the case of methane in slit pore carbons, we have shown that the optimum temperature is about 254 K, given the typical standard enthalpy change of about -16 kJ/mole. Our simulations for methane delivery, depicted in Figure 3 (a), confirmed this result. While the optimal temperature decreases with increase in pore width, as seen in Figure 3 (a), for the pore width of 1.08 nm, which is representative of the modal pore width in most activated carbons, the optimal temperature is about 253 K. At this pore width the maximum absolute delivery of 15.2 mole/kg, or 24.3 wt%, consistent with the estimate of 28.1 wt% maximum delivery at the optimal condition [18]. At larger pore widths the maximum delivery does increase, but at the cost of lower optimal temperature.

Figure 3 (b) depicts the absolute methane delivery as a function of temperature, for carbon nanotubes of different sizes, obtained from our atomistic simulations considering both endohedral and exohedral adsorption for tubes placed in a square array and spaced 0.9 nm apart. The optimum temperature is about 233 K for the largest nanotube examined, having 1.36 nm diameter, and decreases to about 213 K for the three other smaller sizes. These temperatures are lower than the value of 254 K established here for a typical activated carbon, and attained for a homogeneous carbon having 1 nm pores, predominantly due to the heterogeneity of the interstitial space in which the exohedral adsorption occurs. Further, the maximum deliveries range between 14 and 15 mole/kg, which while comparable to the activated carbon of 1.0 nm, are lower than the maximum deliveries for larger pore width carbons, as seen in Figure 3 (a). These results would suggest that, as in the case of hydrogen, carbon nanotubes have no advantages over activated carbon from the viewpoint of methane delivery.

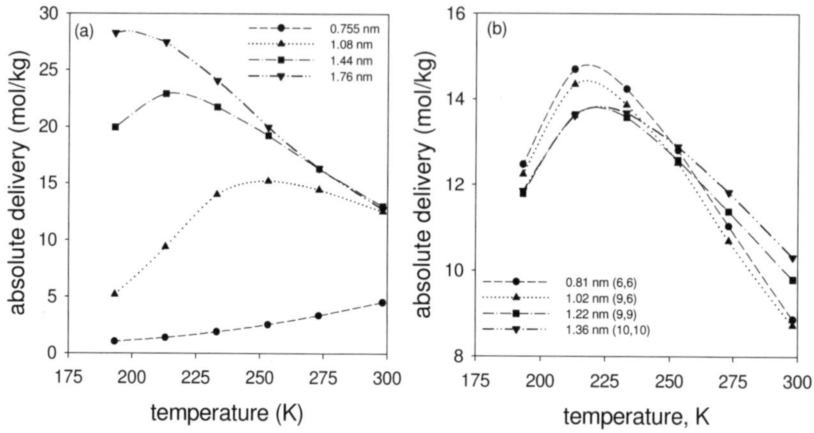

Figure 3. Temperature variation of specific absolute delivery for methane on (a) activated carbons of various pore sizes, and (b) carbon nanotubes of various sizes.

Besides the gravimetric delivery an important measure of the effectiveness of the storage cycle is the enhancement factor, defined as the ratio of delivery from an adsorbent-packed container to that from an identical one filled with bulk gas, operating between 30 and 1.5 bar. To determine this factor we consider a container packed with activated carbon with a bed voidage of 0.26 (the close packed value), and assume the carbon to comprise of macroporosity 0.26, in which the fluid phase density is that of the bulk fluid. Figure 4 (a) depicts the variation of enhancement factor with temperature for hydrogen, for homogeneous carbons of various pore sizes. It is evident that the maximum enhancement factor possible is about 3.1, attained for the 0.9 nm pore width carbon at about 110 K. Thus, the 0.9 nm pore width carbon utilizes the container volume most effectively, though the higher optimal temperature of about 150 K for the 0.755 nm carbon may possibly make this a more attractive option. Nevertheless, it should be noted that the enhancement factors determined here are based on the densest possible packing of spheres, with a void fraction of 26%. In practice the particles will not be spherical but irregular, and lower packing efficiencies will be attained, typically with 30-35% porosity, which will reduce enhancement factors slightly.

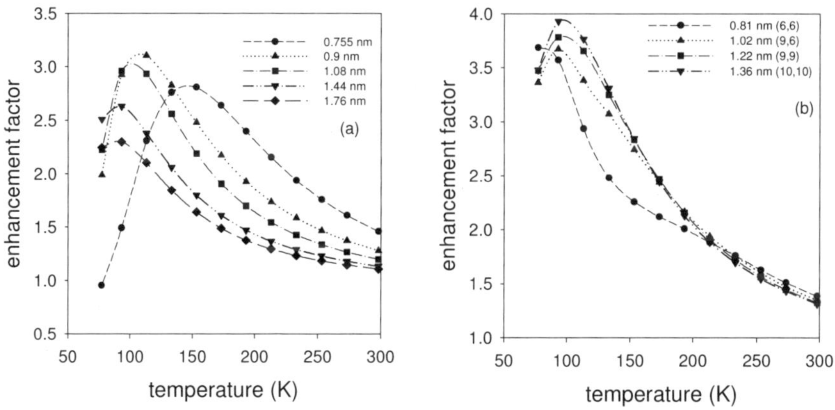

Figure 4. Temperature variation of enhancement factor for hydrogen on (a) activated carbons of various pore sizes, and (b) carbon nanotubes of various sizes.

Figure 4 (b) depicts the enhancement factors for hydrogen storage in carbon nanotube bundles in square geometry In this case an optimal temperature near 100 K is evident for the two largest diameter tubes, with enhancement factors of about 4. However, some reduction in enhancement is likely in comparison to the results in Figure 4 (b) in view of dead spaces created in supporting nanotube bundles in a container, as transport in a fully packed container would be a serious bottleneck for delivery. Nevertheless, the results of Figure 4, showing slightly higher enhancement factors for carbon nanotubes in comparison to slit pore carbons, would suggest that nanotubes can make more efficient utilization of the space. A similar conclusion applies also to delivery of methane from activated carbons and carbon nanotubes [18].

Acknowledgements

This research has been supported by a grant from the Australian Research Council under the Discovery Scheme.

References

1. Huot, J.; Liang, G.; Schulz, R. *Appl. Phys. A* **2001**, *72*, 187.
2. Schlapbach, L.; Züttel, A. *Nature* **2005**, *414*, 353.
3. Vajo, J.J.; Skeith, S.L.; Mertens, F.; Jorgensen, S.W. *J. Alloys Comp.* **2005**, *390*, 55.
4. Vajo, J.J.; Mertens, F.; Ahn, C. C.; Bowman, R.C.; Fultz, B. *J. Phys. Chem.B* **2004**, *108*, 13977.
5. Lee, H.; Lee, J.W.; Kim, D.Y.; Park, J.; Seo, Y.T.; Zeng, H.; Moudrakovski, I.L.;, Ratcliffe, C.I.; Ripmeester, J.A. *Nature* **2005**, *434*, 743.
6. Bénard, P.; Chahine, R. *Langmuir* **2002**, *17*, 1950.
7. Cracknell, R. F. *Phys. Chem. Chem. Phys.* **2001**, *3*, 2091.
8. Schimmel, H.G.; Nijkamp, G.; Kearley, G.J.; Rivera, A.; de Jong, K.P.; Mulder, F.M. *Mat. Sci. Eng.* **2004**, *B108*, 124.
9. Panella, B.; Hirscher, M.; Roth, S. *Carbon* **2005**, *43*, 2209.
10. Takagi H, Hatori H, Soneda Y, Yoshizawa N, Yamada Y *Mat. Sci. Eng.* **2004**, *B108*, 143.
11. van den Berg AWC, Bromley ST, Jansen JC *Micr. Mes. Mat.* **2005**, *78*, 63.
12. Rosi, N.L.; et al. *Science* **2003**, *300*, 1127.
13. Düren, T.; Sarkisov. L.; Yaghi, O.; Snurr, R. *AIChE J.* **2004**, *20*, 2683.
14. Matranga, K.R.; Myers, A.L.; Glandt, E.D. *Chem. Eng. Sci.***1992**, *47*, 1569.
15. Chen, P.; Wu, X.; Lin, J.; Tan, K.L. *Science* **1999**, *285*, 91.
16. Hirscher, M.; Becher, M.; Haluska, M.; Quintel, A.; Skakalova, V.; Choi, Y.M.; Dettlaff-Weglikowska, U.; Roth, S.;, Stepanek, I.; Bernier, P.; Leonhardt, A.; Fink, J. *J. Alloys Comp.* **2002**, *330*, 654.
17. Atsumi, H.; Tauchi, K. *J. Alloys Comp.* **2003**, *356*, 705.
18. Bhatia, S.K.; Myers, A.L. *Langmuir* **2006,** *22*, 1688.
19. Steele, W.A. *The Interaction of Gases with Solid Surfaces*, Pergamon Press, New York, 1974.

Part C: Application

DESULFURIZATION OF FUELS BY SELECTIVE ADSORPTION FOR ULTRA-CLEAN FUELS

YOUN-SANG BAE, JUN-MI KWON AND CHANG-HA LEE
Department of Chemical Engineering, Yonsei University, 134 Shinchon-dong, Seodaemun-gu, Seoul, 120-749, Korea
E-mail: leech@yonsei.ac.kr

Recently, desulfurization for clean-fuel production has gained great interest because of the severe environment regulations and the needs in fuel cell application. The hydrodesulfurization (HDS) process is highly efficient in the desulfurization of liquid fuels. However, it is difficult to use this HDS technology to reduce the sulfur content of liquid fuels to less than 10 ppmw. The new challenge is to use adsorption to selectively remove the sulfur or nitrogen compounds from fossil fuels. There is an ongoing effort to develop new sorbents to remove these compounds in the refinery processes and commercial fuels. In this paper, the desulfurization and denitrogenation by adsorption technology will be reviewed.

1. Introduction

Ultra-deep desulfurization from transportation fuels, particularly from gasoline and diesel, has become very important in petroleum refining industry worldwide not only because of the heightened interest for cleaner air and thus increasingly stringent environmental regulations for fuel sulfur concentration, but also due to the great importance for making ultra-low-sulfur fuels for fuel cell applications [1]. In 1998, the EU first mandated new sulfur specifications for drastically reduced levels that started to be phased from the year of 2000. Similar regulations were legislated in the U.S. and elsewhere soon after. The EPA Tier II regulations request reductions of sulfur in gasoline from 350 to 30 ppmw by January 2005, and those in diesel from the current average of 500 to 15 ppmw by June 2006 [2]. Near future, the regulations plan to be more tightened.

In addition, some fuel cells will require deep-desulfurized fuels. For example, methanol-based fuels for on-board fuel cell applications require the use of a fuel with sulfur content <1 ppmw in order to avoid poisoning and deactivation of the reformer catalyst. To use gasoline or diesel commercial fuels, which are the ideal fuels for fuel cells because of their high energy density, ready availability, and safety and ease for storage, the sulfur concentration should be preferably below 0.1–0.2 ppmw [2].

The hydrodesulfurization (HDS) process is highly efficient in the desulfurization of liquid fuels. However, it is hard to use only the HDS technology to reduce the sulfur content of fuels to less than 10 ppmw, partly because the remaining sulfur compounds in current commercial fuels are thiophenic sulfur compounds which are relatively difficult to remove. Furthermore, the use of amount of sour crude in refinery industries is increasing due to the decrease in natural resource. The technology requires an enhanced catalyst or increased reactor size and/or more severe operating conditions such as high H_2 pressure and high temperature to produce low-sulfur fuels. In the case of gasoline, the need to maintain the octane number by preserving the olefin during HDS makes it more difficult to reach ultra-deep desulfurization to below 5 ppmw in view point of current technology and operating cost [1].

The new challenge is to use adsorption to selectively remove these sulfur compounds from fossil fuels. Since adsorption would be accomplished at ambient pressure and temperature, success in this development would lead to a major advance in petroleum refining. However, success would depend on the development of a highly selective sorbent with a high sulfur capacity, because the commercial sorbents are not adequate for this application [2]. There is an ongoing effort to develop new sorbents to remove the thiophenic compounds from commercial fuels either via π-complexation [2-6], van der Waals' and electrostatic interactions [7,8], and reactive adsorption by chemisorption at elevated temperatures [9,10] among many others.

The aim of this paper is to review the desulfurization by adsorption technology. Before that, we'll briefly introduce the classification of desulfurization technologies.

2. Classification of Desulfurization Technologies

Desulfurization processes can be classified in two groups, 'HDS based' and 'non-HDS based', based on the role of hydrogen in removing sulfur (Table 1). In HDS based processes, hydrogen is used to decompose organosulfur compounds and remove sulfur from refinery streams. However, non-HDS based processes do not require hydrogen [11]. The adsorption is one of the interested strategies among the non-HDS based desulfurization technology.

Table 1. Classification of desulfurization processes

Type	Example	
Catalysis based HDS technology	-Conventional HDS	
	-HDS with fuel specification recovery	
	-HDS by advanced reactor design	
	-HDS by advanced catalysis	
Non-HDS based desulfurization technology	**-Adsorption**	
	-Catalytic distillation	
	-Alkylation	-Extraction
	-Precipitation	-Oxidation

3. Desulfurization by Adsorption

Desulfurization by adsorption is based on the ability of a sorbent to selectively adsorb sulfur compounds from fossil fuels. Based on the mechanism of the sulfur compound interaction with the sorbent, it can be divided into two groups: 'adsorptive desulfurization' and 'reactive adsorption desulfurization'. Adsorptive desulfurization employs physical adsorption of sulfur compounds on the sorbent surface. Regeneration of the sorbent is usually performed by flushing the spent sorbent with a desorbent, resulting in a high sulfur compound concentration flow. Reactive adsorption desulfurization is based on chemical interaction of the sulfur compounds and the sorbent. Sulfur is fixed in the sorbent, usually as sulfide, and the S-free hydrocarbon is released into the purified fuel stream. Regeneration of the spent sorbent results in sulfur elimination as H2S, S, SOx, or sulfur-compounds depending on the process applied [11].

Efficiency of the desulfurization is mainly determined by the sorbent properties: its adsorption capacity, selectivity for the sulfur compounds, durability and regenerability [11]. There has been an ongoing effort to develop new sorbents to remove the sulfur compounds from liquid fuels as summarized in Table 2.

During the past decade, several results have been published on the use of adsorption for liquid fuel desulfurization. Commercially available sorbents (i.e., zeolites, activated carbon, and activated alumina) were used in these studies [7,8,12-14]. However, it is reported that currently available commercial sorbents are not suitable for the adsorptive desulfurization [3].

Initial results on sorbents based on π-complexation for desulfurization were reported by Yang and coworkers and showed these sorbents to be superior to all previously reported sorbents in this application. For desulfurization, they used transition-metal ion exchanged zeolites to selectively remove organo-sulfur molecules from commercial diesel and gasoline [2-6].

Table 2. Studies on the desulfurization by adsorption

Ref.	Sorbents	Treated fuels	Remarkable results
[7]	Activated carbon, Zeolite 5A, Zeolite 13X	Naphtha (550 ppmw S)	Zeolite 13X as well as activated carbon showed much higher sorption capacities for S compounds.
[8]	Activated carbon Zeolites CoMo catalysis Silica-alumina sorbents	Mid-distillate stream (1200 ppmw S)	Activated carbon showed good desulfurization performance at 100°C.
[12]	Zeolites Activated carbon Activated alumina	Thiophene, Benzene	Thiophene adsorbed more selectively than benzene on ZSM-5.
[13]	ZSM-5	Thiophene, Toluene, p-Xylene	Thiophene adsorbed more selectively than toluene and p-xylene on ZSM-5.
[14]	Carbon aerogels	Model diesel (DBT, 4,6-DMDBT)	Carbon aerogels showed good adsorption capacity for both DBT and 4,6-DMDBT.
[15]	Metals Metal halides Metal oxides Metal sulfides Modifies zeolites	Model gasoline (400 ppm S)	Among several types of adsorbents explored, Ni-based adsorbents exhibited better performance for removing sulfur compounds.
[16]	Transition metal-based sorbent	Commercial diesel, gasoline, and jet fuel	Organic sulfur compounds in gasoline, diesel, and jet fuel can be removed by the sorbent.
[3]	Zeolites, Activated carbon, Modified activated Alumina	Thiophene, Benzene	The sorbent capacities for thiophene at the low pressure: Cu-Y, Ag-Y >> Na-ZAM-5 > activated carbon > Na-Y > modified alumina, H-USY.
[4]	Zeolites	Thiophene, Benzene, n-Octane	The sorbent capacities for thiophene: Cu-Y > H-Y > Na-Y > Ag-Y.
[1]	Cu(I)-Y, Ni-based sorbent	Commercial gasoline (305 ppmw S)	Cu(I)-Y and Ni-based adsorbent showed the sorbent capacities of 0.22 and 0.37 mg S/g of sorbent at room temperature, respectively.
[5]	Cu(I)-Y, Ag-Y	Commercial diesel (430 ppmw S) Commercial gasoline	The sulfur content was reduced from 430 to <0.2 ppmw at a sorbent capacity of 34 cm^3 of clean diesel produced per g of sorbent.
[6]	Cu(I)-Y γ-Al_2O_3/Cu(I)-Y	Commercial diesel (297.2 ppmw S)	The γ-Al_2O_3/Cu(I)-Y showed the desulfurization capacity of 0.29 mmol S/g of zeolite.
[2]	Cu(I)-Y	Diesel, Gasoline, Jet fuel	The sorbent capacities of 0.395 and 0.278 mmol S/g of sorbent for jet fuel and diesel, respectively.

Ma and coworkers recently synthesized various adsorbents including metals, metal halides, metal oxides, metal sulfides, and modified zeolites and evaluated their desulfurizing abilities in their laboratory. Their approach aims at removing sulfur compounds in gasoline and jet fuels selectively by a direct sulfur-adsorbent interaction, rather than π-complexation [1,10,15,16].

In the conventional HDS process, refractory sulfur-contining compounds (SCCs) are deprived of the chance to take up the active sites to be hydrogenated because of the higher adsorptivity of nitrogen-containing compounds (NCCs). Therefore, if these NCCs are effectively removed from liquid fuels prior to the HDS process, the limitation of HDS process can be overcome (Fig. 1).

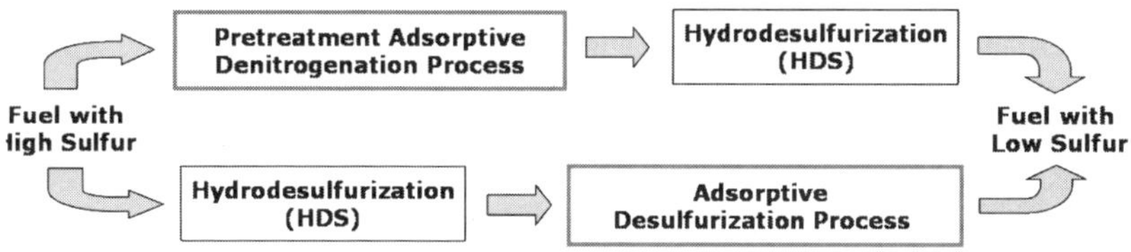

Figure 1. Pretreatment adsorptive denitrogentation and direct adsorptive desulfurization processes for the ultra deep desulfurization of fuels.

Table 3. Studies on the pretreatment adsorptive denitrogenation

Ref.	Sorbents	Treated fuels	Remarkable results
[17]	Cu(I)-Y	Commercial diesel (83 ppmw N)	Cu(I)-Y showed sorbent capacity of 3 mg N/g sorbent.
[18]	Si-Zr cogel	Light gas oil (190 ppmw N, 8200 ppmw S)	Si-Zr cogel exhibited sorbent capacity of 4.7 mg N/g sorbent.

Some studies have been performed to develop new sorbents to remove the nitrogen compounds from fossil fuels prior to HDS process (Table 3). Hernandez-Maldonado and Yang [17] showed that Cu(I)-Y zeolite can effectively remove nitrogen from a commercial diesel fuel that contains 83 ppmw nitrogen to well below 0.1 ppmw nitrogen at a sorbent capacity of 43 cm^3 diesel per g of sorbent. This corresponds to a very high and practical sorbent capacity of 3mg N/g sorbent.

Recently, the sorption characteristics of NCCs on the Si-Zr cogel were reported for the denitrogenation of light gas oil (LGO) by our group [18]. The LGO contained NCCs of about 190 ppmw and SCCs of about 8,200 ppmw. The saturated sorption capacity of the Si-Zr cogel was about 4.7 mg N/g sorbent at 50°C. In addition, the ability of desorption and re-adsorption of NCCs was studied by using three kinds of solvents (MTBE, MIBK, and Anisole). Now, our group synthesized several novel adsorbents to directly remove NCCs and SCCs from fossil fuels and the ability of selective adsorption is superior to the present adsorbents.

Acknowledgements

Financial support from the Korean Ministry of Environment as "The Eco-technopia 21 Project" is gratefully acknowledged.

References

1. Ma X., Velu S., Kim J.H. and Song C., *Appl. Catal. B* **56** (2005) pp. 137-147.
2. Hernandez-Maldonado A.J. Yang F.H., Qi G. and Yang R.T., *Applied Catalysis B: Environmental* **56** (2005) pp. 111-126.
3. Takahashi A., Yang F.H. and Yang R.T., *Ind. Eng. Chem. Res.* **41** (2002) pp. 2487-2496.
4. Hernandez-Maldonado A.J. and Yang R.T., *Ind. Eng. Chem. Res.* **42** (2003) pp. 123-129.
5. Yang R.T., Hernandez-Maldonado A.J. and Yang F.H., *Science* **301** (2003) pp. 79-81.
6. Hernandez-Maldonado A.J. and Yang R.T., *J. Am. Chem. Soc.* **126** (2004) pp. 992-993.
7. Salem A.B.S.H. and Hamid H.S., *Chem. Engng. Technol.* **20** (1997) pp. 342.
8. Savage D.W., Kaul B.K., Dupre G.D., O'Bara J.T., Wales W.E. and Ho T.C., US patent 5,454,933.
9. Khare G.P., US Patent 6184176 (2001), to Phillips Petroleum Company.
10. Velu S., Ma X., Song C., *Ind. Eng. Chem. Res.* **42** (2003) pp. 5293.
11. Babich I.V. and Moulijn J.A., *Fuel* **82** (2003) pp. 607-631.
12. Weitkamp J., Schwark M. and Ernest S., J. *Chem. Soc. Chem. Commun.* (1991) pp. 1133.
13. King D.L., Faz C. and Flynn T., *SAE paper 2000-01-0002, Society of automotive engineers*: Detroit, MI, 2000.
14. Jayne D., Zhang Y., Haji S. and Erkey C. International Journal of Hydrogen Energy *30* (2005) pp. 1287-1293.
15. Ma X., Sprague, M., Sun L. and Song C., *Am. Chem. Soc. Div. Fuel Chem. Prepr.* **47** (2002) pp. 452.
16. Ma X., Sun L. and Song C., *Catal. Today* **77** (2002) pp. 107-116.
17. Hernandez-Maldonado A.J. and Yang R.T., *Angew. Chem. Ind. Ed.* **43** (2004) pp. 1004-1006.
18. Bae Y.-S., Kim M.-B., Lee H.-J., Min W.S. and Lee C.-H., *AIChE J.* **52** (2006) pp. 510-521.

LARGE SCALE CO SEPARATION BY VPSA USING CUCL/ZEOLITE ADSORBENT

Y. C. XIE, J. ZHANG, Y. GENG, W. TANG AND X. Z. Tong
State Key Lab for Structural Chemistry
College of Chemistry, Peking University
Pioneer Technology Company
Beijing 100871, China,
E-mail: yxie@pku.edu.cn

Based on the principles that cuprous ions can form complex with carbon monoxide and salts can spontaneously disperse onto the surface of supports as a monolayer, a highly efficient CO adsorbent, CuCl/Zeoilte, has been made by heating a mixture of CuCl and a zeolite at a suitable temperature to disperse the CuCl onto the surface of the zeolite. This adsorbent has high CO adsorption capacity (>50ml/g at 1 atm. and ambient temperature) and high CO selectivity over H_2, N_2, CH_4 and CO。 Using this adsorbent in a VPSA process, a large scale plant has been designed and built in China for separation of CO from syngas. The feed gas contains about CO 30%, H_2 41%, N_2 17%, CO_2 8%, CH_4 2%, O_2 0.4%,and saturated water. The plant can produce carbon monoxide 1700m^3 per hour with purity >99% and recover >85%.

1. Introduction

Carbon monoxide is an important raw material in chemical industry. It can be used for the synthesis of many chemicals, such as acetic acid, acetic anhydride, formic acid, dimethyl carbonate, polycarbonate, N,N-dimethylformamide (DMF), oxalates, propinoic acid, acrylic acid phosgene, polyisocyanates (TDI and MDI), polyurethanes and metal carbonyls etc. There are many sources of CO in industry, such as synthesis gas from steam reforming and partial oxidation of natural gas, oil and coal as well as by-product gases from steel and iron plants or other industries. In these gases, carbon monoxide is coexistence with N_2, H_2, CO_2, CH_4 and H_2O etc. The separation of CO from the gas mixtures is of great interest in industries. Conventional way to separating CO from gas mixture is cryogenic process[1]. The process needs pretreatment, using liquid absorbent such as MEA, DEA or MEDA to remove bulk CO_2 and a thermal swing adsorption to remove water and trace CO_2 at first, and then uses cryogenic distillation at low temperature and high pressure to obtain pure CO. The process is high energy consumption and its equipments is high cost. An absorption process named COSOB, which used $CuCl.AlCl_3$ in toluene as absorbent, had been developed to separating CO by Tenneco company in 1970's [1], but it had been superseded in industry application owing to the serious corrosion problem.

Adsorbents and pressure swing adsorption (PSA) process for separation of CO has been developed by many labs and companies [2-8]. Although some commercial technologies to separate CO by PSA have been reported, they have not been adopted widely in industry owing to that their adsorbents have CO capacity and selectivity not good enough or cause corrosion problem. We have developed and patented highly efficient CO adsorbents before [9]. With a highly efficient CO adsorbent, CuCl/zeolite, and a reasonable VPSA process (Pressure Swing Adsorption with Vacuation), a large scale plant has been designed and built in China to produced CO from syngas. This plant has been operating continually and smoothly for three years to produce high purity CO with high recovery. The properties of the adsorbent and performance of the plant are reported in this paper.

2. Highly efficient CO adsorbent

Common adsorbents, such as activated carbon, silica gel, alumina and zeolites, are not suitable for CO separation from gas mixtures containing N_2, H_2, CO_2, CH_4 and H_2O, because they have low adsorption capacity and selectivity for CO. It is well known that cuprous ion (Cu^+) can form complex with CO, if a great amount of cuprous compound is put on the surface of a support with high surface area, it is possible to get an adsorbent with high CO adsorption capacity and selectivity. In our fundamental research work, it has been found that many oxides and salts can disperse spontaneously onto the surface of supports to form a monolayer [10]. Based on this principle, we mixed CuCl and a zeolite and heated them at a suitable temperature, the CuCl can disperse onto the surface of the zeolites as a monolayer, so an adsorbent with very high capacity and selectivity for CO was obtained [11-14]. Using this technology, a highly efficient CO adsorbent, named PU-1, has been commercialized by Pioneer Technology Company in China.

Figure 1 shows the adsorption isotherms of CO, CO_2, CH_4, N_2 and H_2 for the adsorbent at ambient temperature. The adsorbent adsorbs CO much more than H_2, N_2 and CH_4, showing that the adsorbent has high CO adsorption capacity and selectivity over CH_4, N_2 and H_2. The adsorbent adsorbs CO also more than CO_2, though the CO selectivity over CO_2 is not as great as CH_4, N_2 and H_2.

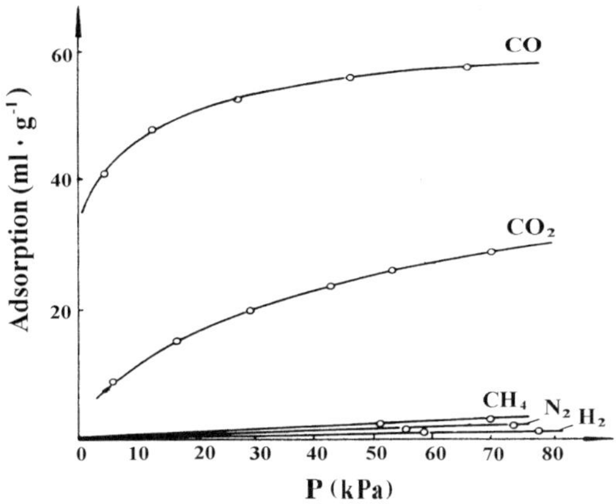

Figure 1. Adsorption isotherms of CO, CO_2, CH_4, N_2 and H_2 for PU-1 adsorbent at 20 °C

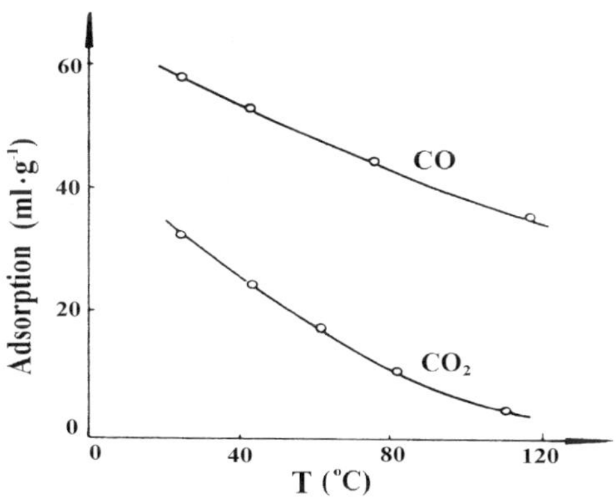

Figure 2. Adsorption isobars of CO and CO_2 for PU-1 adsorbent at 450 mmHg

When temperature is increase, the adsorption capacity of CO_2 on PU-1 declines much faster than CO as shown in Figure 2. It indicates that the CO selectivity over CO_2 can be improved by raising temperature.

Figure 3 shows the CO breakthrough curve of a gas mixture of CO and N_2 for the PU-1 adsorbent. Before the breakthrough point, the CO concentration in the effluent is lower than 5 ppm. It shows that the adsorbent has very good performance for CO separation from N_2, which is very difficult in cryogenic process.

Figure 4 shows the breakthrough curve of CO and CH_4 of a gas mixture of CO, CH_4 and H_2 for the adsorbent. It shows that the separation of CO from CH_4 and H_2 is also very good. Methane is a very harmful impurity in CO for the production of phosgene, TDI, MDI and polycarbonate. The very good performance of the adsorbent for the separation between CH_4 and CO is important for the production of CO for these products.

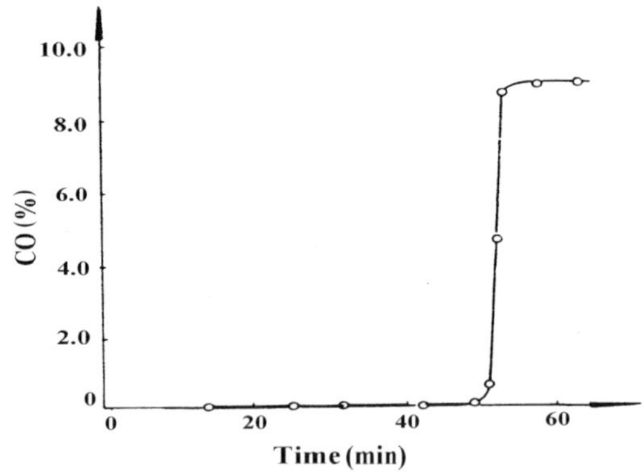

Figure 3. Breakthrough curve of CO for PU-1 adsorbent at space velocity 500 ml/hr.g. Feed composition 9.0 % CO and 91 % N_2. 20 °C, 1 atm.. The fluent contains CO < 5 ppm before the breakthrough point.

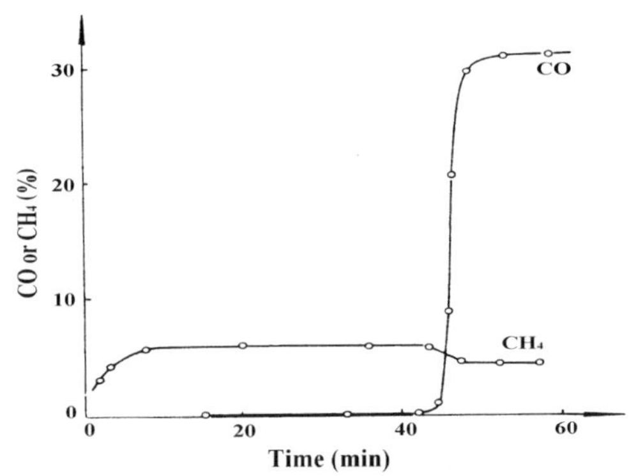

Figure 4. Breakthrough curves of CO and CH_4 for PU-1 adsorbent at space velocity 200 ml/g.hr. Feed composition 4.0 % CH_4, 30.7 % CO, 65.3% H_2. 15 °C, 1 atm..
The effluent contains CO < 5 ppm before breakthrough point.

3. Plant for CO separation with VPSA processes

By using the PU-1 adsorbent, a large scale plant using VPSA processes (Pressure Swing Adsorption with Vacuation) has been designed and built in China to produce 1700 m^3/hr CO from syngas for production of acetic anhydride. The plant consists of two units, a pre-treatment VPSA-1 unit to remove CO_2, water and trace heavy components such as sulfur-containing compounds, followed by a VPSA-2 unit to produce CO.

The first unit VPSA-1 has three adsorber filled with adsorbents which have high adsorption capacity for CO_2 and H_2O and poor adsorption for CO. The feed composition is about 30% CO, 41% H_2, 17% N_2, 8%CO_2, 2.2% CH_4, 0.4% O_2 and saturated water. It is compressed to about 8 atm. at room temperature before feeding to the VPSA-1 unit. Each adsorber passes through the following steps in cycle: adsorption, pressure-equalization, counter depressurization, purge with tail gas from VPSA-2 and evacuation (regeneration), partially pressurization (with gas from pressure equalization), re-pressurization (with purified gas from adsorption step). The cycle time of the VPSA-1 is about 20mins. The effluent from VPSA-1 contains CO_2<100ppm and H_2O<100ppm is used as the feed of VPSA-2.

The second unit VPSA-2 has four adsorber filled with PU-1 adsorbent.. A schematic of the four bed VPSA process for CO separation is shown in Figure 5.

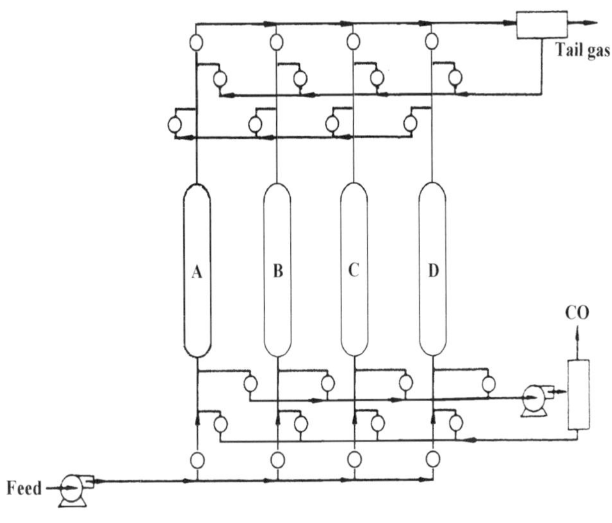

Figure 5. Schematic of the four bed VPSA process

In stead of room temperature the adsorbers are operated at about 70°C in order to increase the working capacity of CO and CO selectivity over CO_2. Each adsorber passes through the following steps in cycle:

a) Adsorption (Ad): The feed gas is fed under about 7.5 atm. through the adsorber until CO is just beginning to breakthrough. The adsorbed phase is primarily CO, and the tail gas mainly consists of other gases (H_2, N_2, CH_4 and CO_2). At the end of the feed step, the void gas composition in the adsorber is essentially the feed composition.

b) Pressure equalization (PE): The gas in the adsorber is co-current expansion to another adsorber which just finishes the evacuation step (step 5) to start the pressure build up step (step 6). The pressure in the two adsorber becomes equal and about half of the feed pressure. This step can decrease the loss of CO.

c) Purge (Pu): In order to remove the impurity gas co-adsorbed on the adsorbent and remained in the void space of the adsorber, the adsorber is purged with a part of CO product at an intermediate pressure. This step is responsible for the high purity of the CO product obtained at the next two steps (steps 4 and 5). The purity of the CO product can be controlled by the quantity of the purge. Effluent from this step goes into another adsorber which just finishes the pressure build up step (step 6). Some residual gas may flow out from the another adsorber. The residual gas might be compressed and recycled to the feed to increase CO recovery.

d) Depressurization (Dep): After the purge step, the adsorber is counter depressurization to atmosphere for desorption and recovery of CO as product.

e) Evacuation (Ev): The adsober is evacuated for further desorption of CO from the adsorbent to obtain high purity CO product.

f) Pressre build-up (PBu): After the evacuation step, the adsober is pressure build-up with expansion gas from another adsorber at pressure equalization step (step 2).

g) Pre-loading (PL): The adsorber receives effluent from another adsorber which is at purge step (step 2).

h) Repressurization (ReP): The adsorber is connected to another adsorber which is undergoing adsorption step. This step repressurizes the adsorber to adsorption pressure and makes it available for the adsorption step (step 1) of next cycle.

Each adsorber undergoes the above cyclic steps in a sequential manner. The four adsorbers are operated in turn to make the process works continuously. All these are achieved by opening and closing the suitable valves connecting to the adsorbers according to a time program which is controled by a computer. Table 1 shows the sequence of cyclic process steps of the four adsorbers. The time period for each step has been tested and found to obtain the best result. The cycle time is about 12 minutes.

The fluent from VPSA-1 was heated to about 70°C to feed to VPSA-2 at about 7.5 atmosphere. In VPSA-2 process, the evacuating pressure is 0.15-0.20 atmosphere , purge pressure about 3 atmosphere, the purge ratio is about 0.3, the cycle time is about 12 minutes. The plant has obtained the following results: CO product 1700 m^3/hour, CO recovery >85%, purity >99%, impurity CH_4<188ppm, CO_2<10ppm, O_2<5ppm. The plant has commissioned in Feb. 2003 in China and has been operating continually and smoothly in good condition until now.

Table 1. Process steps of VPSA -2 for CO separation.

Bed						Steps*						
A	Ad	Ad	Ad	PE	Pur	Dep	Ev	Ev	Ev	PBu	PL	ReP
B	PBu	PL	ReP	Ad	Ad	Ad	PE	Pur	Dep	Ev	Ev	Ev
C	Ev	Ev	Ev	PBu	PL	ReP	Ad	Ad	Ad	PE	Pur	Dep
D	PE	Pur	Dep	Ev	Ev	Ev	PBu	PL	ReP	Ad	Ad	Ad

*Ad, Adsorption; PE, Pressure equalization; Pur, Purge; Dep, Depressurization; Ev, Evacuation; PBu, Pressure build-up; PL, Pre-loading; ReP , Re-pressurization

4. Conclusion

A highly efficient CO adsorbent has been obtained by heating a mixture of CuCl and a zeolite at a suitable temperature. This adsorbent has high adsorption capacity and selectivity for CO over H_2, N_2, CH_4, and CO_2. Using this adsorbent in a VPSA process, a large scale CO separation has been succeeded in obtaining CO with purity >99% and recovery >85% from a syngas gas containing about 30% CO and rich in N_2 , CH_4 and CO_2.

Acknowledgments

The authors acknowledge the supports by The Major Basic Research Development Program (Grant No. G 2000077503) and by National Science Foundation of China (NSFC).

References

1. Haddeland G. E., SRI International Report, No.123 Carbon Monoxide Recovery, 1979.
2. Hirai H., .Wada K., and Komigama M., Chemistry Letter, 261 (1983) .
3. Benkmann C., Linde Report on Science and Technology, No.44, p.8, (1988).
4. Tajima K., and Osada Y., Nippon Konan Technical Report, Oversea, No.50 (1987); U.S. Patent 4,783,433 to Nippon Kokan Kabushiki Kaisha (1988).
5. Yokoe J., Takeuchi M., Tsuji T., U.S. Patent 4,713,090(1987) to Kansai Netsukagaka Kabushiki Kaisha.
6. Golden T. C., Kratz W. C., and Withelm F.C., U.S. Patent 5,126,310 to Air Products and Chemicals, Inc. (1992).
7. Kumar R., Kratz W.C., Guro D.E. and Golden T.C., Separation Technology, edited by E.F.Vansant,1994 Elsevier B.V. p.383-402.
8. Golden T.C., Guro D.E., Kratz W.C., Occhialini J.M. and Sabram T.E., Fundamentals of Adsorption 6, (Elsevier,1998, Francis Meunier ad.), 695.
9. Xie X. Y., Bu N., Liu J.., Yang G., Qiu J. G., Yang N. F., Tang, Y. Q., U.S. Patent, 4,917,711(1990); Canada Patent 1304343, 1992..
10. Xie Y, C., Tang Y. Q., Advances in Catalysis, Vol.37.1 (1990).
11. Xie Y. C.,, Yang G., Qiu J. G., Tong X. Z., Liu J., Luo,B., Tang Y. Q., Fundamentals of Adsorption, M Suzuki Ed., Kodansha, 737(1993) .
12. Xie Y.C., Zhang J.P., Qiu J. G., Tong X.. Z., Fu J. P., Yang G., Yan H.J., Tang Y.Q, Adsorption, 3, 27 (1996).
13. Xie Y. C., Zhang J. P., Tong X. Z., Pan X.. M., Fu J. P., Cai X.H., Yang G. and Tang Y. Q., Chemical Journal of Chinese Universities, Vol.18, 7, 1159(1997).
14. Zhang J. P., Pan X.M., Fu J. P., Long X. Y., Qiu J. G., Cai X. H., Xie Y. C., Tang Y. Q., Fundamentals of Adsorption 6, F. Meunier Ed., Elsevier (1998).

THE ZLC METHOD FOR DIFFUSION MEASUREMENTS

STEFANO BRANDANI

Department of Chemical Engineering, University College London, Torrington Place, London

WC1E 7JE, UK

E-mail: s.brandani@ucl.ac.uk

The zero length column (ZLC) technique has become a common tool to measure mass transfer kinetics in microporous adsorbents. The partial loading experiment is a variant of the traditional ZLC method in which the adsorbent is not allowed to reach full equilibration with the gas phase. Even though this variant of the ZLC experiment was introduced over 10 years ago, it has been applied only by few researchers. In this contribution we review the basic theory of the partial loading experiment and show that it can be used to establish the contributions of different mass transfer mechanisms. A detailed numerical model that includes the effects of nonlinearity of the isotherm and combined diffusion and surface barrier effects is presented to allow the correlation of complex sorbate-sorbent systems.

1. Introduction

The ZLC method was introduced by Eic and Ruthven [1] in the late eighties and has now become a standard technique to measure mass transfer kinetics in porous materials. The normal technique consists of a very short chromatographic column that is initially equilibrated with a stream containing the adsorbate. At time zero the inlet valve is switched and a stream of pure carrier is used to desorb the adsorbate. This is repeated at different flowrates and provided that the system is far from equilibrium control the mass transfer kinetics are determined using the solution to the diffusion equation applied to a perfectly mixed cell [1, 2].

The solution to the diffusion equation yields a series of exponentials and it is difficult from a single ZLC experiment to distinguish different mass transfer mechanisms, i.e. surface barriers vs internal diffusion. For linear systems the shape of the initial part of the desorption curves should be distinctive [3] and the analysis of the moments of the desorption curves can also provide a means to distinguish the two mechanisms [4].

Both these methods are not applicable to nonlinear systems and Brandani and Ruthven [5] introduced the partial loading experiment in order to have a further means to distinguish between diffusion and surface barriers. In this case the system is exposed to the adsorbate/carrier gas mixture for a limited time in order to load only in part the adsorbent material. Therefore in a partial loading experiment if the mass transfer mechanism is due to diffusion, when the inlet valve is switched the solid will have an internal concentration profile, while if the system is controlled by a surface barrier the concentration inside the particle will be uniform and similar to the fully equilibrated case. Evaluating from a mass balance the average adsorbed phase concentration it is therefore possible to distinguish clearly the two mechanisms [5]. In this contribution we review the general theory and present a model that includes the effect of system nonlinearities.

2. Theory

The basic idea behind the ZLC experiment is to maximize axial dispersion in a chromatographic column by reducing the length. Therefore the mass balance equation can be formulated in terms of the kinetics of a perfectly mixed cell [2]:

$$V_S \frac{d\bar{q}}{dt} + V_F \frac{dc}{dt} = F\left(c_{IN} - c\right) \tag{1}$$

where c is the gas phase concentration; c_{IN} is the inlet concentration; F is the volumetric flowrate; $\bar{q}$ is the average adsorbed phase concentration; t is time; V_F is the volume of the fluid and V_S is the volume of the solid.

To include the effect of isotherm nonlinearity and limit the number of additional parameters we will consider for simplicity that the Langmuir equation can represent the adsorption equilibrium:

$$q^* = q_S \frac{bc}{1+bc} \tag{2}$$

where the Henry law constant $K = bq_S$; q^* is the equilibrium concentration and q_S is the adsorbate concentration at saturation.

The mass balance in the cell, eq. (1), is coupled to the mass balance in the solid by:

$$V_S \frac{d\bar{q}}{dt} = -S_S J\big|_{R_P} \tag{3}$$

where J is the molar flux and S_S is the surface of the solid. Assuming the presence of both a surface barrier and internal diffusion

$$J\big|_{R_P} = -k\left(q^* - q_{R_P}\right) = -D_0\Gamma \frac{\partial q}{\partial r}\bigg|_{R_P} \tag{4}$$

where D_0 is the corrected diffusivity; k is the mass transfer constant and Γ is the thermodynamic correction factor for the diffusion coefficient [6]. For simplicity we will assume both k and D_0 to be independent of concentration. From eq. (2) and the definition of Γ [6]

$$\Gamma = \frac{q_S}{q_S - q} \tag{5}$$

The mass balance in the solid completes the set of equations for the model:

$$\frac{\partial q}{\partial t} = \frac{1}{r^{\sigma-1}}\frac{\partial}{\partial r}\left(D_0\Gamma r^{\sigma-1}\frac{\partial q}{\partial r}\right) \tag{6}$$

where σ depends on the geometry of the adsorbent material: 1 for a slab; 2 for a cylinder and 3 for a sphere.

The model equations can be rewritten in terms of dimensionless variables

$$\xi = \frac{r}{R_P} \qquad \tau = \frac{D_0 t}{R_P^2} \qquad C = \frac{c}{c_0} \qquad Q = \frac{q}{q_0}$$

and parameters

$$\beta = \frac{V_F c_0}{\sigma V_S q_0} \qquad L = \frac{F R_P^2 c_0}{\sigma D_0 V_S q_0} \qquad \lambda = \frac{q_0}{q_S} \qquad \delta = \frac{k R_P}{D_0}$$

In terms of the equivalent parameters in the case of a linear isotherm

$$\beta_0 = \frac{V_F}{\sigma V_S K} \qquad L_0 = \frac{F R_P^2}{\sigma D_0 V_S K}$$

the following hold for a Langmuir isotherm [7]

$$\frac{q_0}{c_0} = K(1-\lambda) \qquad \beta = \frac{\beta_0}{1-\lambda} \qquad L = \frac{L_0}{1-\lambda}$$

In dimensionless form, eqs (1-2) and (4-6) become:

$$\Gamma\frac{\partial Q}{\partial \xi}\bigg|_1 + \beta\frac{dC}{d\tau} = L\left(C_{IN} - C\right) \tag{7}$$

$$Q^* = \frac{C}{1-\lambda+\lambda C} \tag{8}$$

$$\Gamma \frac{\partial Q}{\partial \xi}\bigg|_1 = \delta\left(Q^* - Q_1\right) \tag{9}$$

$$\Gamma = \frac{1}{1-\lambda Q} \tag{10}$$

$$\frac{\partial Q}{\partial \tau} = \frac{1}{\xi^{\sigma-1}} \frac{\partial}{\partial \xi}\left(\Gamma \xi^{\sigma-1} \frac{\partial Q}{\partial \xi}\right) \tag{11}$$

For gaseous systems the parameter β is typically less than 0.1 and the accumulation in the fluid phase can be neglected. In the actual solution this term will be retained with $\beta_0 = 0.01$ since it stabilizes the numerical integration.

The partial loading experiment can be performed only if an internal concentration profile can be generated, i.e. if the system is far from equilibrium control. This can be achieved if the parameter L_0 is greater than 10. The parameter L is directly linked to the internal concentration gradient and this can be seen from eq. 7, since at time zero when the valve is switched C = 0 and $C_{IN} = 1$:

$$\frac{\partial Q}{\partial \xi}\bigg|_1 = \frac{L}{\Gamma} = L_0 \tag{12}$$

Note that the final equality holds only for a Langmuir isotherm. If L_0 is too small it will not be possible to generate an internal concentration gradient, since the gradient at the surface at time zero is the maximum gradient in the particle at any time.

The partial loading experiment introduces a new parameter τ_{PL}, which is the dimensionless load time which can be varied easily. In the analysis the valve dynamics will be assumed to be much faster than the diffusional and surface barrier time constants and the inlet concentration will be represented as a square wave. In the experiment only the gas phase concentration is measured, but a simple mass balance can be used to obtain the adsorbed phase concentration

$$\overline{Q}_{PL} - \overline{Q} + \sigma\beta\left(C_{PL} - C\right) = \sigma L \int_{\tau_{PL}}^{\tau} C\, d\tau \tag{13}$$

where $\overline{Q}_{PL}$ is the average adsorbed phase concentration at the end of the loading step.

3. Diffusion control: δ >> 1

The general model described in the theory section reduces to the diffusion control limit if the mass transfer resistance introduced by the surface barrier can be neglected, i.e. $\delta \gg 1$. In order to have a qualitative understanding of the effect of a partial loading experiment we will consider $L_0 = 20$ and vary τ_{PL} and λ and fix $\delta = 100$. Figure 1 shows the results of the simulations for $\lambda = 0.1$, 0.5 and 0.9.

These cases are representative of a linear system, a mildly nonlinear system and a strongly nonlinear system. As can be seen from the results, the nonlinearity has the effect of shifting the long time asymptotic decay.

The effect of partial loading in a linear system can be seen for $\tau_{PL} < 0.25$, while for the nonlinear systems the loading time needs to be reduced due to the thermodynamic correction factor that increases the diffusivity.

Figure 2 shows the adsorbed phase concentrations corresponding to the previous cases. In the case of internal diffusion, the solid phase concentrations are dependent upon the loading times, τ_{PL}.

4. Surface barrier control: δ << 1

The general model described in the theory section reduces to the surface barrier control limit if the diffusional time constant is small compared to that of the surface barrier, i.e. $\delta \ll 1$. In order to have a qualitative understanding of the effect of a partial loading experiment we will consider $L_0 = 20$ and vary τ_{PL} and λ and fix $\delta = 0.1$.

Figure 3 shows the results of the simulations for $\lambda = 0.1$ and 0.9. These cases are representative of a linear system and a strongly nonlinear system. As can be seen from the results, the nonlinearity has the effect of shifting the long time asymptotic decay. Note that in the gas phase plot also for the surface barrier controlled system there is a shift resulting from the decreasing loading times. Qualitatively this is the same result as for diffusion control.

From Figure 4 it is evident that the adsorbed phase concentration plots are independent of the loading time and can be used to distinguish the two mass transfer mechanisms. The nonlinearity of the system does not have any influence on this result.

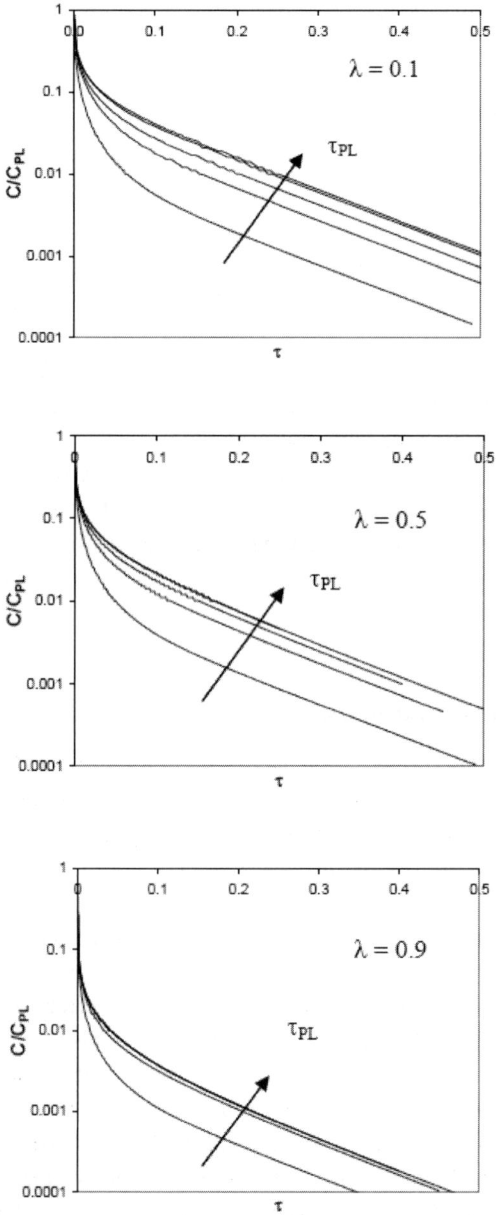

Figure 1. Gas phase concentrations normalized at τ_{PL} = 0.5, 0.25, 0.1, 0.05 and 0.01

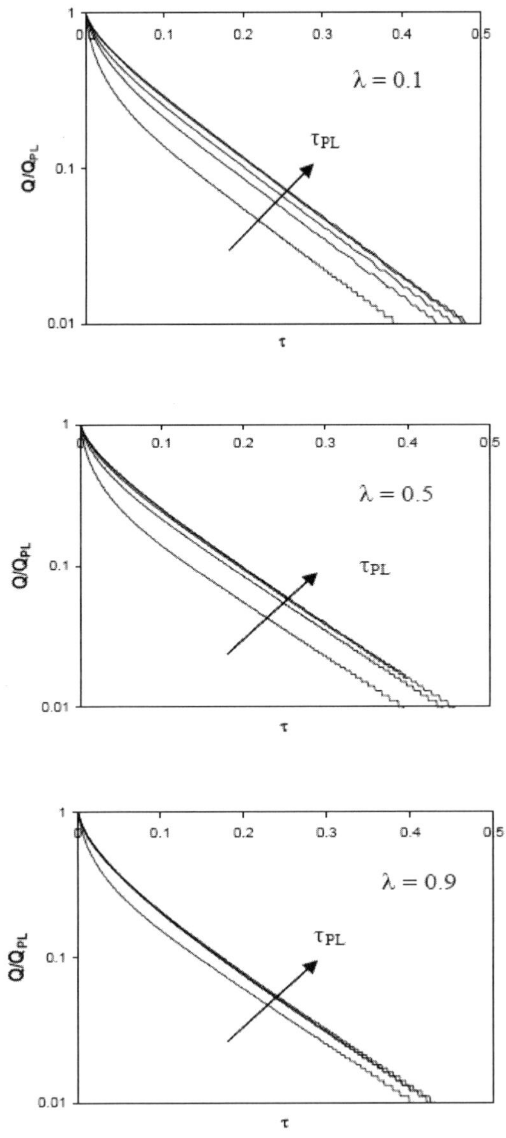

Figure 2. Adsorbed phase concentrations normalized at τ_{PL} = 0.5, 0.25, 0.1, 0.05 and 0.01

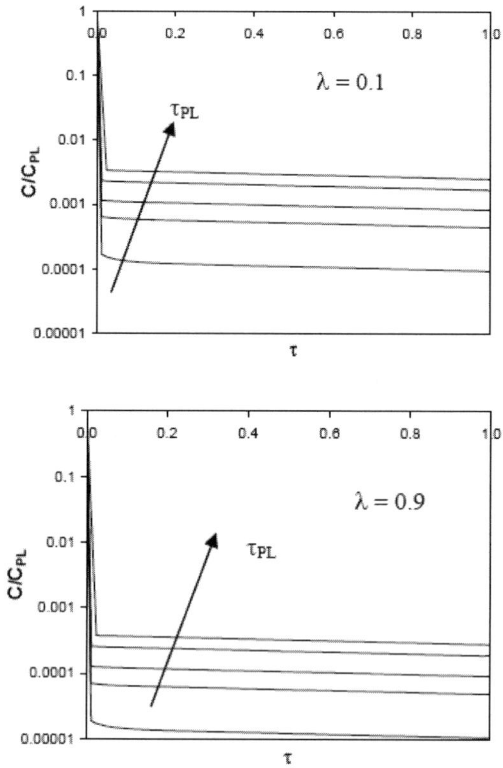

Figure 3. Gas phase concentrations normalized at τ_{PL} = 5, 2.5, 1, 0.5 and 0.1

5. Discussion

The ZLC partial loading experiment can be used to distinguish clearly the limiting mass transfer mechanisms of internal diffusion and surface barriers. This approach can be applied with confidence to both linear and nonlinear systems and provides a simple way to generate multiple ZLC response curves that can be used to extract kinetic information.

To fully characterize an adsorbate-adsorbent system one should run ZLC experiments at low flowrates to obtain the adsorption isotherm [8]. Having obtained the isotherm, possibly also through other independent measurements, one should use a numerical code to obtain the limiting diffusivity and surface barrier kinetic constant from the simultaneous fit of multiple high flowrate and partial loading experiments. It should be noted that the partial loading experiment, together with experiments at multiple flowrates, can be used also to show that the system is not under equilibrium control.

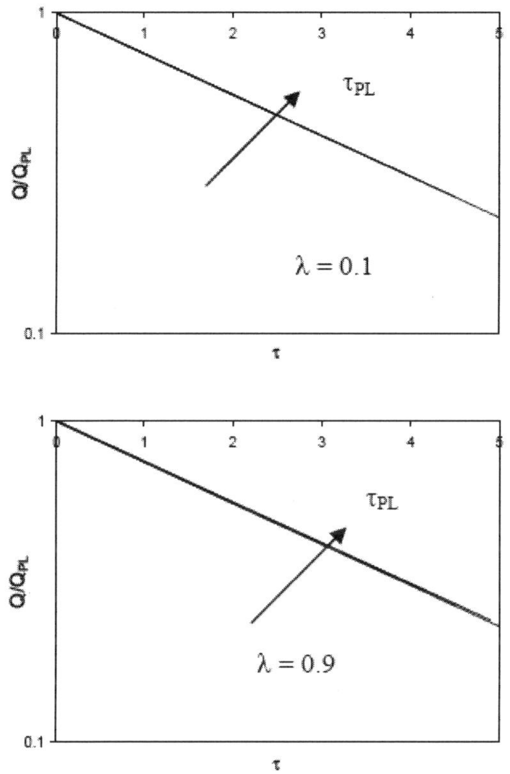

Figure 4. Adsorbed phase concentrations normalized at τ_{PL} = 5, 2.5, 1, 0.5 and 0.1

Acknowledgements

The discussions with the partners of the International Research Group on "Diffusion in Zeolites" (http://www.uni-leipzig.de/diffusion/pages/irg.html) have been one of the motivations for this contribution. This work was carried out in part while at UOP Ltd on an industrial secondment sponsored by UOP and the Royal Academy of Engineering. Financial support from the EPSRC (GR/R95142/01) and the Royal Society Wolfson Research Merit Award is gratefully acknowledged.

References

1. Eic M. and Ruthven D. M., A new experimental technique for measurement of intracrystalline diffusivity. *Zeolites* **8** (1988) pp. 40–45.
2. Brandani S. and Ruthven D. M., Analysis of ZLC desorption curves for liquid systems. *Chem. Eng. Sci.* **50** (1995) pp. 2055–2059.
3. Ruthven D. M. and Brandani F., ZLC response for systems with surface resistance control. *Adsorption* **11** (2005) pp. 31–34.
4. Brandani S. and Ruthven D. M., Moments analysis of the zero length column method. *Ind. Eng. Chem. Res.* **35** (1996) pp. 315–319.
5. Brandani S. and Ruthven D. M., Analysis of ZLC desorption curves for gaseous systems. *Adsorption* **2** (1996) pp. 133–143.
6. Ruthven D. M. Principles of adsorption and adsorption processes (Wiley, New York, 1984).
7. Brandani S., Effects of nonlinear equilibrium on zero length column experiments. *Chem. Eng. Sci.* **53** (1998) pp. 2791–2798.
8. Brandani F., Ruthven D. M. and Coe C. G., Measurement of adsorption equilibrium by the zero length column (ZLC) technique part 1: single-component systems. *Ind. Eng. Chem. Res.* **42** (2003) pp. 1451–1461.

CHIRAL SEPARATION OF PROPRANOLOL HYDROCHLORIDE BY SMB PROCESS INTEGRATED WITH CRYSTALLIZATION

XIN WANG, YUE LIU AND CHI BUN CHING

Division of Chemical and Biomolecular Engineering School of Chemical and Biomedical Engineering Nanyang Technological University Singapore 637722
E-mail: xwang@ntu.edu.sg

Resolution of propranolol hydrochloride was studied in self-packed columns of perphenyl carbamoylated beta-cyclodextrin (beta-CD). Both bed voidage and linear equilibrium constants were evaluated from a series of linear elution chromatograms by moment analysis. A modified h-root method was used to determine the competitive Langmuir isotherm of propranolol hydrochloride in the nonlinear region. Continuous separation of the target enantiomer from its racemic mixture was studied by Simulated Moving Bed (SMB) chromatography in both linear and nonlinear region. Desired (S)-propranolol hydrochloride was produced in the raffinate product at a high purity. Solubility of propranolol hydrochloride was determined experimentally in methanol at different temperatures. Crystallization of propranolol hydrochloride from different initial composition solutions in the mixed solvent of methanol and acetone was also investigated with different product purity and yield. SMB productivity was further increased at the sacrifice of decreasing product purity. The obtained solution was further purified by crystallization process. Compared with direct crystallization which is only suitable for racemic conglomerate, the integrated process is especially suitable for the majority of chiral drugs which belong to racemic compounds as long as suitable and economic chiral stationary phases (CSPs) are available in the SMB separation.

1. Introduction

The chirality of drugs is an important issue from pharmacological, pharmacokinetic, toxicological and regulatory points of view [1-2]. Nowadays more research efforts have been concentrated on the production of optically pure products due to increasing demand that such drugs are administered in optically pure form [3]. Propranolol belongs to the most important beta-blocker drugs since a variety of analogous compounds have been developed based on it. It is mainly used in the treatment of hypertension and cardiac arrhythmias and it has been reported that its desired activity resides in the S-(-)-enantiomer form. Propranolol hydrochloride has one chiral center and is supplied in its hydrochloride from, as shown in Figure 1.

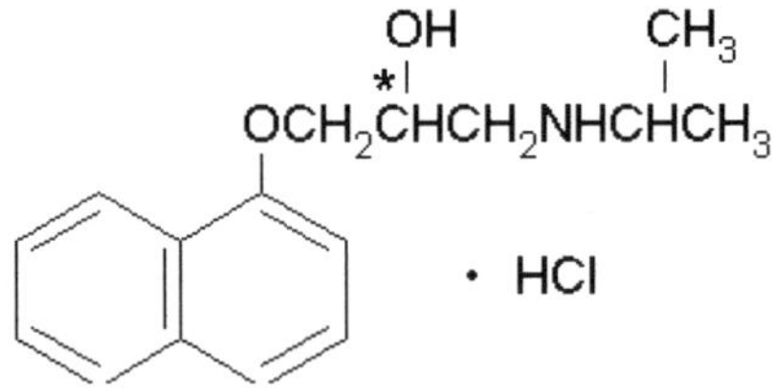

Figure 1. Molecular structure of propranolol hydrochloride

Simulated Moving Bed (SMB) process has been extensively applied to the separation of chiral drugs and intermediates over the last decade [4-7]. Due to continuous countercurrent contact between liquid and solid phases, SMB process allows the decrease of desorbent requirement and the improvement of productivity per unit time and unit mass of stationary phase. SMB process is believed to be able to achieve high purity separation even when the resolution exhibited by an individual column is not efficient for a batch preparative process, which is often the case in chiral separations. One of the key issues in operating SMB process is to determine zone flow rates and column switching time. Developed in the frame of equilibrium theory which neglects the effect of axial mixing and mass transfer resistances, triangle theory are currently widely applied SMB design approaches [8-9]. In this method, development of SMB is resort to its corresponding hypothetical true counter-current (TCC) process and the most important parameters required are those of the bed voidage (or total porosity) and equilibrium isotherms of the enantiomers to be separated. The TCC operation parameters can then be converted to SMB unit based on the geometric and kinematic equivalence between the two processes [10-11].

However, the high cost of the enantioseparation process, especially the chiral stationary phases (CSPs) which usually demonstrate good enantioseparation abilities towards specific compounds/drugs, makes the large-scale application of SMB in chiral separation less favourable. Crystallization technique on the other hand remains an important and economic process for industrial-scale production and purification of enantiomers [12]. Racemate crystals can be divided into racemic compound, racemic conglomerates and pseudoracemates (solid solutions). Although diastereomer crystallization, which is often referred to as classical resolution, has been studied in detail for more than a hundred years, the selection of resolving agent is still a matter of trial and error. Preferential crystallization is more attractive but can only be directly accomplished for conglomerates. Unfortunately, only 5-10% of all racemates are conglomerates, the majority of chiral substances belong to racemic compound. Only partially resolved solution enantioriched by other technique, whose composition is over the eutectic composition, can be separated by this technique.

The coupling of liquid chromatography, especially SMB process and crystallization has been investigated recently for efficient enantioseparation [13-15]. In this study, resolution of racemate propranolol hydrochloride was achieved on a column packed with perphenyl carbamoylated β-cyclodextrin (β-CD) immobilized onto silica gel. Both bed voidage and linear equilibrium constants were evaluated from a series of linear elution chromatograms conducted at different interstitial velocity. A modified h-root method was used to determine the competitive Langmuir isotherm of propranolol hydrochloride in the nonlinear region. Complete separation of racemic mixture of propranolol hydrochloride by SMB was achieved in both linear and nonlinear regions. The solubility of racemate and enantiomer of propranolol hydrochloride in the solvent of methanol was determined experimentally at different temperatures. Crystallization of propranolol hydrochloride from different initial composition solutions in the mixed solvent of methanol and acetone was investigated with different product purity and yield. To increase the productivity of the desired (S)-enantiomer, SMB experiment was run at higher feed concentration and zone flow rates with partially resolved product obtained in the raffinate stream. The obtained solution were concentrated and purified by crystallization process.

2. Theoretical Background

2.1. *Column physicochemical properties and adsorption isotherm*

The bed voidage can be evaluated from the zero retention time of a non-adsorbed component to the stationary phase. For a component which enters the pore system but does not adsorb on the surface of the stationary phase, the retention time of such a component is given by:

$$t_{OR} = \frac{V\varepsilon_T}{\dot{V}} = \frac{L\varepsilon_T}{u} \tag{1}$$

For packing materials with two pore systems of micropores and macropores, the column total porosity ε_T and bed voidage ε can be related by equation:

$$\varepsilon_T = 0.45 + 0.55\varepsilon \tag{2}$$

It is well known that chromatographic separation depends primarily on the adsorption isotherms, which relates the solutes concentration in the mobile phase to that of the stationary phase over the concentration range of interest. In the diluted region, linear isotherm was expressed as:

$$q_i^* = K_i \cdot C_i \tag{3}$$

The method of moments is used to determine the adsorption equilibrium of the column. For a linear isotherm model, the first moment is expressed as [16]:

$$\mu_1 = \frac{L}{v}\left[1+\left(\frac{1-\varepsilon}{\varepsilon}\right)K\right] \tag{4}$$

The first moments of the enantiomers to be separated can be plotted against the inverse interstitial velocity of mobile phase and linear equilibrium constants can be readily determined from the slopes of the lines.

It is well known that SMB is preferably conducted in nonlinear region to achieve higher productivity; therefore it is more important to determine the competitive adsorption behavior among the feed species. In special, the non-stoichiometric Langmuir isotherm is important in SMB development since constraints on the flow rate ratios (i.e., m_1, m_2, m_3 and m_4) in SMB unit can be determined explicitly on the frame of equilibrium theory [8]. It can be expressed as:

$$q_j^* = \frac{a_j c_j}{1+\sum_{i=1}^{n} b_i c_i} \tag{5}$$

where a_i are measures of the intrinsic affinities of the respective species for the sorbent, and the b_i are characteristic of the nature and strength of interference produced by the species. It is worth noticing that the linear isotherm can be seen as particular case of the nonstoichiometric Langmuir and linear equilibrium constants K_i is equal to Langmuir coefficients a_i.

The h-root method without the introduction of dummy species has been applied to determine the non-linear competitive Langmuir isotherms of nadolol, a three chiral center beta-blocker drug [17]. In this method, the individual isomers of interest, which are often not commercially available, are not required and only very small amount of racemic mixture is needed. This facilitates the determination of isotherms for racemic drugs. This method divides the determination of Langmuir parameters into two parts. The intrinsic affinity coefficients a_i were obtained from linear elution chromatography, and competitive interference coefficients b_i were obtained from non-linear frontal chromatography.

The equations used to determine the competitive Langmuir coefficients of racemic mixture are given as follows [18-19]:

$$\sum_{i=1}^{n}\left(\frac{c_i^f}{\frac{k_i^{'}}{K_n^{'}}-1}b_i\right)=1 \tag{6}$$

$$\sum_{i=1}^{n}\left(\frac{c_i^f}{\frac{k_i^{'}K_{j+1}^{'}}{k_{j+1}^{'}K_j^{'}}-1}b_i\right)=1 \quad j=1,2,\cdots n-1 \tag{7}$$

where C_i^f are feed concentrations, $k_i^{'}$ and $K_i^{'}$ are elution capacity factors and frontal capacity factors, respectively.

In equations (6) and (7) all the terms are known or can be experimentally determined, except that of the Langmuir competitive adsorption coefficients b_i. Thus n equations can be used to determine the unknown b_i $(i=1,2,\cdots n)$.

2.2. *SMB separation of propranolol hydrochloride*

In the frame of equilibrium theory, which neglects mass transfer resistances and axial dispersion, true counter-current (TCC) adsorption model was employed in a series of efforts to obtain explicit expressions of the fluid to solid flow rate ratios, m_j $(j=1,\cdots 4)$, for complete separation of binary mixtures [8-9, 20-23]. The operation condition of SMB was then determined based on the equivalence between SMB and TCC process by keeping constant the liquid velocity relative to the solid velocity in the two processes. In special, desorbent is usually nonadsorbable (or it is so weak that its adsorptivity is negligible) for enantiomeric separation, and explicit criteria were obtained [8] to determine the boundaries of the complete separation region in the space spanned by m_j $(j=1,\cdots 4)$. It should be noted that the purity and yield of both components are 100 % in theory within the complete separation region.

Fluid phase flow rate over solid phase flow rate of TCC unit can be defined as:

$$m_j = \frac{Q_j^{TCC}}{Q_S} = \frac{v_L \varepsilon}{v_S (1-\varepsilon)} \tag{8}$$

which can be converted to the flow rate ratios of the equivalent SMB unit using the conversion equation:

$$m_j = \frac{Q_j^{SMB} t^* - V\varepsilon}{V(1-\varepsilon)} \tag{9}$$

The parameters m_j (j=1,...4) define a four-dimensional space divided into different regions, and it is useful to consider the projection of the four-dimensional regions onto (m_2, m_3) plane. The boundaries between the different separation regions depend only on the adsorption isotherm of the mixtures to be separated and feed concentration and composition. Having decided m_j (j=1,...4) and t^* (or Q_I), Equation 9 is often used to determine the liquid flow rate in the four sections of SMB and thus the inlet & outlet streams flow rates. The advantage of this approach is that the flow rate ratio is a dimensionless group bringing together information about column volume, V, unit flow rates, Q_i, and switch time, t^*, and thus can be applied whatever the configuration, size and productivity of the SMB unit in both linear and non-linear systems.

3. Experimental

3.1. *Chemicals*

HPLC-grade methanol was obtained from Fisher Scientific (Leics, UK). Glacial acetic acid and triethylamine were obtained from Merck (Germany). HPLC water was made in the laboratory using a Millipore ultra-pure water system. The racemate mixture of propranolol hydrochloride was purchased from Sigma (St. Louis, MO, USA). All purchased products are used without further purification.

Empty column (25 cm x 1 cm I. D.) assembly was purchased from Phenomenex (USA). The columns were packed with perphenyl carbamoylated beta-cyclodextrin bonded onto silica gel using an Alltech pneumatic liquid pump (Alltech, USA) by slurry packing method. The silica gel was supplied by Eka Chemicals AB (Sweden) with particle size of 16 μm (KR100-16-SIL). The eluent (desorbent) used was a binary mixture containing 60% aqueous buffer solution (1% TEAA, pH=4.5) and 40% methanol. The feed solution was prepared by dissolving racemate propranolol hydrochloride in the desorbent at certain concentrations. The eluent and feed solution were degassed in a model LC 60H ultrasonic bath before running the experiment.

3.2. *SMB separation system*

In the SMB unit, the countercurrent contact between the solid and mobile phase is achieved by the periodically shifting the inlet (feed, desorbent) and outlet (raffinate, extract) ports in the direction of the fluid flow. In this work, the SMB separation unit is open-looped and consists of 8 columns (25 cm x 1 cm I. D.) arranged in a 2-2-2-2 configuration, i.e., two columns per section (see Figure 2). Five flows (feed, eluent, extract, raffinate, and recycled eluent) are needed to handle in the SMB unit. The flow rates of two inlet streams, i.e., feed and eluent, as well as two of the three outlet streams, e.g., extract and raffinate, are controlled and thus leaving the recycled eluent stream free and determined by the overall material balance of the SMB unit. An online vacuum degasser (SUPELCO) degasses all the liquid being pumped into the system.

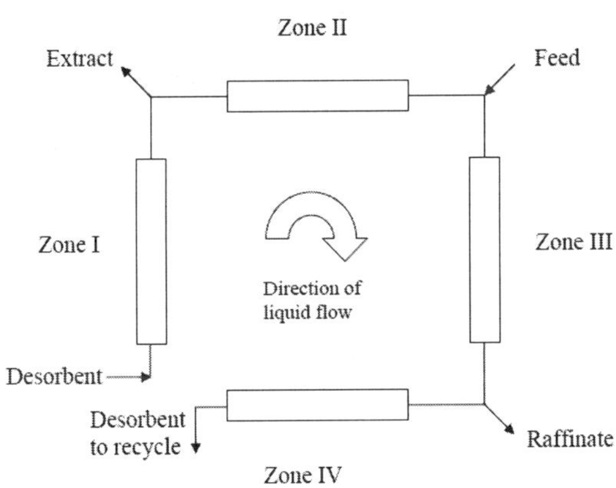

Figure 2. Schematic diagram of SMB unit: 8 columns, 2-2-2-2 configuration, open looped

The concentrations of the extract and raffinate streams were analyzed using Shimadzu SCL-10AVP chromatographic system. The samples of products were collected at the middle of the switch times at different cycle and switch times. An analytical column (25 cm x 0.46 cm I. D.) packed by perphenyl carbamoylated β-CD bonded onto 5μm silica gel was used to analyze the concentration of samples based on calibration lines obtained previously from external standard solutions. The absorbance wavelength was set at 220 nm. All chromatographic experiments were conducted at room temperature around 23 °C.

4. Results and Discussions

4.1. *Elution order of the enantiomers of propranolol hydrochloride*

In order to determine the elution order of enantiomers of propranolol hydrochloride, samples of the two stereoisomers of propranolol hydrochloride were injected into the column respectively under the same chromatographic conditions as that for the racemic mixture of propranolol hydrochloride. It was found that (S)- and (R)- propranolol hydrochloride correspond to the first and second peak of racemate propranolol hydrochloride, respectively. Thus (S) - and (R) - propranolol hydrochloride are enriched in the raffinate and extract streams in the SMB experiments, respectively.

4.2. *Determination of bed voidage*

1,3,5 tri-tert-butyl benzene (TTBB) has been widely used for the determination of column dead time t_{OR} for various CSPs [24]. Although the sorption to the perphenyl carbamoylated β -cyclodextrin is strongly supported by a phenyl group, this group is surrounded and shielded by the three tert-butyl groups in the case of TTBB. Further more, an exclusion mechanism is not likely to occur due to the relatively small molecular size of TTBB. Therefore, TTBB is believed not to be retained in the stationary phase and was chosen to determine the total porosity ε_T of the column in this study.

The total porosity ε_T, was determined from the response to a pulse injection of TTBB. The retention time of TTBB in the column was corrected by deducting the retention time of TTBB peak measured when the injector directly connected to the detector.

The zero retention time of TTBB was given by Equation 1. From the plot of mean retention time against the inverse flow rate in Figure 3, the total porosity ε_T was found to be 0.64. From Equation 2, the bed voidage was found to be 0.34 for the column.

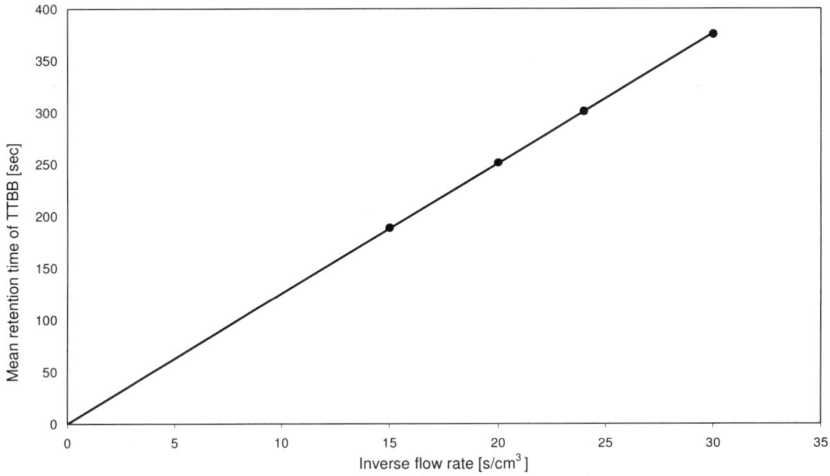

Figure 3. Plot of mean retention time of TTBB against mobile phase inverse flow rate

4.3. *Determination of equilibrium isotherm*

The linear isotherm was valid only in linear concentration range. Thus all pulse experiments need to be carried out under dilute conditions. Dilute propranolol hydrochloride samples were used in the chromatographic experiment and with continuous decreasing of the amount of samples injected, there were only very slight difference for the first moments of the two peaks. According to the experimental results, concentration of propranolol hydrochloride solution at 0.104 mg/ml is believed to be in the linear isotherm region.

The first moments of the two components of propranolol hydrochloride were plotted against the inverse superficial velocity of mobile phase in Figure 4. Straight lines were fitted to the experimental points. According to Equation 4, the equilibrium constants were determined from the slopes of the lines, which were found to be 4.36 and 6.31 for (S)-propranolol hydrochloride and *(R)*-propranolol hydrochloride, respectively.

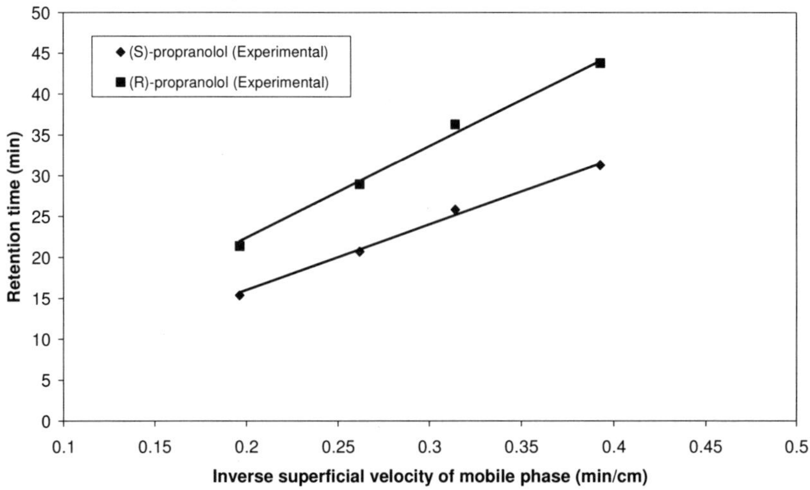

Figure 4. Retention time of propranolol hydrochloride versus inverse superficial velocity of mobile phase

The h-root method without the introduction of dummy species was applied to determine the non-linear competitive Langmuir isotherms of the two enantiomers. Although ideally only one frontal experiment is necessary to determine the competitive Langmuir coefficients b_i, the possibility of experimental error and the difficulty to determine T_i accurately necessitates other confirming frontal experiments, which may be conducted at different concentrations of the step changes of the solutes and at different flow rate of the mobile phase. In this study, the experiments were conducted at concentrations of propranolol hydrochloride at 0.754 mg/ml and 1.004 mg/ml, respectively and the flow rate of the mobile phase was 3 ml/min and 4 ml/min, respectively.

The competitive Langmuir coefficients of the two components of propranolol hydrochloride were evaluated at the average of b_1 and b_2 and the final isotherms at the concentration range studied were given as:

$$q_1^* = \frac{4.357c_1}{1+1.484c_1+3.495c_2}$$

$$q_2^* = \frac{6.307c_2}{1+1.484c_1+3.495c_2}$$

4.4. *SMB separation of propranolol hydrochloride*

In the design of SMB experiments, one is mostly concerned with the projection of the four-dimensional space, m_j (j=1,...4), onto (m_2, m_3) plane, i.e., the plane in the operating parameter space spanned by the flow rate ratios of the two key sections of the SMB unit. From adsorption isotherm determined previously and the feed concentration, complete separation regions for propranolol hydrochloride separation was constructed in the (m_2, m_3) plane, as shown in Figure 5. It is worth noting that for proper operation of SMB to obtain desired complete separation, adsorbent and fluid should be regenerated in section 1 and 4 respectively.

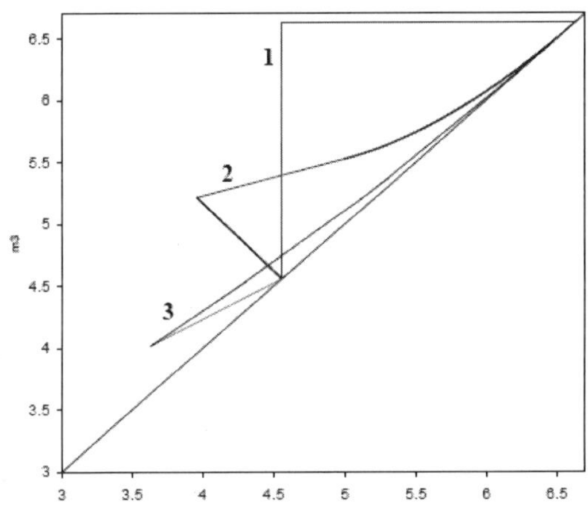

Figure 5. Different separation regions in SMB experiments. Feed concentration: ((1)-0.15 mg/ml; (2)-0.75 mg/ml; (3)-1.5 mg/ml)

At the SMB's theoretical optimum operating state, the unit has the highest possible productivity and enrichment of products and the lowest desorbent consumption. However, the performance of the SMB at this condition is not robust and is very sensitive to various kinds of disturbances.

Basically, the SMB operation points should be close to the theoretical optimal point in order to achieve a high production rate, yet far away from it within the boundaries of the operating area to assure robustness. Since (S) - propranolol hydrochloride is the desired enantiomer product which is enriched in the raffinate stream, productivity based on raffinate rather than on the feed to

SMB is more useful. From Equation 9, raffinate productivity based on unit CSP volume can be deduced as follows:

$$P_{Raf} = \frac{c_B^R Q_R}{(1-\varepsilon)VN_C} = \frac{c_B^R (m_3 - m_4)}{t^* N_C} \tag{10}$$

In order to increase raffinate productivity, one can either increase the difference of $(m_3 - m_4)$ or decrease the switching time. Various SMB experiments were run at different operation conditions. The operating parameters and separation performance such as purity and productivity are examined, which are shown in Table 1.

Table 1. Operating conditions and separation results of SMB experiments

Run	Flow rate ratios				Switch time	Flow rates (ml/min)				Product purity (%)		Raffinate Productivity (mg/day)
	m_1	m_2	m_3	m_4	t^*(min)	Q_1	Q_F	Q_R	Q_E	Raf	Ext	
C	7.19	4.76	4.96	3.58	15	6.33	0.16	1.12	1.97	99.9	99.8	15.91
D	7.19	4.92	5.12	3.60	15	6.33	0.16	1.23	1.84	99.6	99.2	17.47
F	7.48	4.60	4.70	2.90	17	6.06	0.07	1.29	2.33	99.6	99.4	39.05
G	7.48	4.79	4.88	2.47	17	6.07	0.07	1.83	2.04	92.5	91.8	34.56
H	7.48	4.67	4.76	2.37	17	6.09	0.07	1.82	2.14	80.2	77.8	55.56

Run C and D were run in the linear isotherm range and m_3 in run D was increased (i.e., the operation condition was changed along the operation line toward the pure extract region). It was found that the product purities in both product streams are nearly 100 %, which is consistent with the complete separation regions. The productivity in Run D is slightly higher since the operation point is moved along operation line in the direction of increasing the difference of $(m_3 - m_4)$. Run F and G were run in nonlinear range at a concentration of 0.754 mg/ml, while $(m_3 - m_2)$ was further increased at Run G with the attempt to increase raffinate productivity. However, only partially resolved products were obtained indicating less robustness of this run. Run H was performed at higher concentration of 1.5 mg/ml, which exceed the concentration range within which the Langmuir isotherm was determined. Raffinate product with the highest productivity and 80 % purity was obtained.

It was found that SMB can separate both enantiomers in high purity, e.g., in Run C and D if operation points were chosen inside the complete separation region and one does not seek high productivity of the desired product. It is also suggested that SMB can be operated to achieve partially separated products of interest with higher productivity. This can be followed by a simple crystallization step to obtain the pure enantiomer.

It is worth noting that some experimental results do not agree well with theoretical predictions. This could stem from different chemico-physical parameters of columns in the SMB unit and the difficulty of controlling flow rates accurately in the SMB experimental studies.

4.5. *Solubility phase diagram of propranolol hydrochloride system*

For the study of crystallization from solution, it is useful to determine the solid/liquid equilibrium solubility diagram of the racemic species of interest. The ternary solubility diagram is helpful to understand the nature of racemic mixture. In fact, the feasibility and yield of enantioseparation of a partially resolved mixture is dependent on the shape of the phase diagram and the position of eutectic points. In consideration of the solvent used in the chromatography separation process, methanol was selected as crystallization solvent in the experiments. The solubility of propranolol hydrochloride in methanol was measured by classical visual-polythermal method and the results are shown in Figure 6. In the polythermal method, solvent and solute are weighed into a small closed glass vessel in suitable proportions. The contents are heated gently with agitation until all of the crystals have been dissolved. The clear solution is first cooled until it nucleates. The temperature is then increased slowly (lower than 0.2 °C/min) until the last crystal dissolves. At this point the equilibrium saturation temperature has been achieved. The procedures are repeated by adding solute or solvent to obtain the solubility data in the desired temperate range.

The ternary solubility phase diagram of (S) - and (R) - propranolol hydrochloride in a mixed solvent of methanol and acetone was measured by isothermal method [25]. For isothermal method, enough amount of powder, namely 100±0.1mg, was dissolved in the solvent of methanol in a test tube. Saturated solution samples were carefully withdrawn and filtered, and the concentration of which were analyzed by the HPLC system with employment of above-mentioned self-packed column.

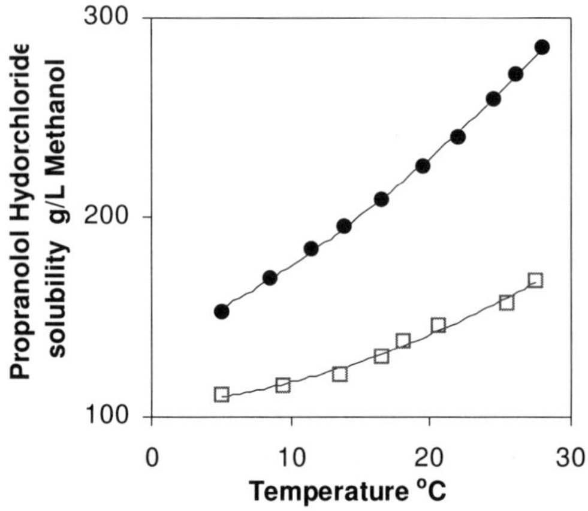

Figure 6. Solubility of propranolol hydrochloride in methanol. ● (R, S) - propranolol hydrochloride; □ (S) - propranolol hydrochloride.

The solubility data helps one to choose the most suitable condition for crystallization operation. In binary chiral system, solubility phase diagram is essential for identifying the region for crystallization resolution. Due to thermodynamic constraint, for almost 95 percent of the chiral substances which belong to racemic compound, crystallization separation is likely to succeed only when the initial solution composition is above the eutectic point. From Figure 6, propranolol hydrochloride is highly soluble in methanol and the solubility data of both (R,S)-and (S)-propranolol hydrochloride in methanol show an obvious increasing trend as the temperature increases and the solubility curve of racemate has a deeper slope than that of enantiomer. Due to stability concern, solubility data higher than 30°C was not determined.

The solid-state properties of propranolol hydrochloride in respect of the relationship between the racemic mixture and (S) - enantiomer have been previously reported [25]. The shape of a ternary phase diagram can theoretically be deduced from respective binary phase diagram. Similar to the results of the binary melting point phase diagram, ternary phase diagram shows a shape of a typical conglomerate type compound [25]. However, the two eutectic points are so close to each other that the exact position of eutectic points is not likely to be determined precisely.

4.6. *Crystallization of propranolol hydrochloride system*

Propranolol hydrochloride was identified as a racemic compound although it possesses the phase diagram of conglomerate shape. The eutectic points are close to the racemic mixture, which means resolution might be successful by crystallization of solution at a low enantiomeric excess (e.e). The favorable temperature range to be identified for the crystallization operation is the region within which solubility of racemate is much higher than that of enantiomer. Crystallization resolution of (R, S) - propranolol hydrochloride was performed under constant temperature of 15°C in 1:2(V/V) methanol and acetone mixture (the mixture of methanol and acetone instead of pure methanol was employed as the crystallization solvent here due to the suitable solubility of propranolol hydrochloride). Dissolving certain quantity of racemate in the solvent at 30°C and then slowly cooling the solution to the desired experimental temperature 15°C, thoroughly collect the crystals and analysis the product purity. Crystallization results are shown in Table 2.

Table 2. Preferential crystallization of (R, S) - propranolol hydrochloride

Run	Initial Quantities (mg)	Initial R:S Ratio	Seed (mg)	Product e.e (%)	Yield (%)
1	300	50:50	15	0	28.2
2	300	65:35	15	78.5	25.5
3	300	70:30	15	90.8	18.6
4	300	75:25	15	91.2	16.7

Preferential crystallization attempts performed on a racemate solution (Run 1) failed to obtain the enantiomer pure product, which might be due to the lower lattice energy for the two enantiomers packed orderly in one single crystal in a racemic compound system. Started from a higher initial purity, for example 70%, relatively high purity crystals were obtained. The 91.2 % product e.e. (Run 4) rather than pure crystals of one enantiomer is due to the difficulty of separation of crystals from the mother liquor. The successful removal of mother liquor is crucial for higher product e.e because the retaining two enantiomers mixture of mother liquor in the crystal product will work as impurities thus decrease the final product purity. In addition to the initial solution purity, the separation process is controlled by another essential factor, the degree of supersaturation. A highly supersaturated solution most likely leads to the deposit of racemate even when seeded with pure enantiomer. On the other hand, a lower supersaturation will suffer the difficulty in increasing the product yield.

4.7. *Crystallization of propranolol hydrochloride from SMB products*

Although the eutectic points of propranolol hydrochloride are close to the racemic mixture, crystallization of racemate solution or solution at a low enantiomeric excess (e.e) failed to get pure enantiomer product. SMB process on the other hand can be operated to produce optically pure enantiomer, e.g., in Run C, D and F at productivity of 15.9, 17.5 and 39 mg/day. Certain amount of solution from SMB Run H was concentrated and crystallized using the method discussed previously, final product of (S) - propranolol hydrochloride with 92.5 % e.e. was obtained. The integrated SMB and crystallization process thus theoretically could give a productivity of 53.5 mg/day (pure (S)-enantiomer), which is higher than that produced by SMB process alone. It should be mentioned that with further increasing of SMB productivity, more crystals can be obtained from crystallization which facilitates the process of washing off mother liquor. This could give a higher e.e product and thus increase the final amount of the desired enantiomer. In the future study, SMB experiments could be performed at higher feed concentration, larger product flow rate and higher enrichment for the desired component. It is worth noting that the solvent selection is difficult and important. It should provide good separation capacity since it is used as mobile phase and deosrbent in batch chromatography and SMB separations respectively. It should also have suitable solubility for the sample of interest since it is also the crystallization solvent. In the future study, the integrated process is to be investigated in normal phase which facilitates the removal of solvent to obtain pure crystal product.

5. Conclusions

Based on column physicochemical properties and adsorption equilibrium isotherm determined, continuous separation of the target enantiomer of propranolol hydrochloride from its racemate mixture was studied by SMB chromatography in both linear and nonlinear region. The solubility of racemate and enantiomer of propranolol hydrochloride in the solvent of methanol was determined experimentally at different temperatures. Crystallization of propranolol hydrochloride from different initial composition solutions in the mixed solvent of methanol and acetone resulted in different product purity and yield. Further, crystallization of the concentrated enantioriched solution from SMB process, the composition of which being above the eutectic point composition, crystals with high purity was obtained. The integrated process is found to be feasible and promising for racemic compound forming chiral system.

Symbols used

a_i Intrinsic affinity coefficients (dimensionless)

b_i Langmuir competitive interference coefficient (ml/mg)

c_i Mobile phase concentration based on fluid volume (mg/ml)

c_i^F Feed concentration (mg/ml)

k' Elution capacity (retention) factor of the solute (dimensionless) calculated from linear elution chromatography ($k_i' = \dfrac{1-\varepsilon}{\varepsilon} \cdot a_i$)

K_i Equilibrium constant (dimensionless)

K_i' Frontal capacity factor (dimensionless) calculated from non-linear frontal chromatography ($K'_i = \dfrac{T_i - T_0}{T_0}$)

L Column length (cm)

m_j Fluid phase flow rate over sold phase flow rate in j section of TCC and SMB unit

N_C Total number of columns in SMB

q_i Concentration of component i on stationary phase (mg/ml)

q_i^* Equilibrium concentration of component i on stationary phase (mg/ml)

Q_F Feed flow-rate fed to SMB process

Q_j Liquid phase flow rate in *j* section of TCC or SMB process

Q_s Solid phase flow rate in TCC process

t^* Switching time in SMB process (min)

t_{0R} Mean retention time for an unretained compound (min) (when compound can enter the pore system of the stationary phase)

T_0 Column hold up time in frontal experiments (min)

T_i Breakthrough time of the waves in frontal experiments (min)

u Superficial velocity (cm/s)

v Interstitial fluid velocity of the mobile phase (cm/s)

v_L Interstitial fluid velocity of the fluid phase in SMB process

v_s Solid velocity in TCC process

V Column volume

V Volumetric flow rate of the mobile phase (ml/min)

ε Bed voidage

ε_T Total porosity of column

L Liquid phase

S Solid phase

1 The first eluted component of propranolol hydrochloride racemic mixture (component 1 or component B)

2 The second eluted component of propranolol hydrochloride racemic mixture (component 2 or component A)

SMB Simulated moving bed chromatography

TCC true counter-current chromatography

F SMB Feed stream

R SMB raffinate product

E SMB extract product

References

1. R. Nation, *Clin. Pharmacokinet.* 27, 249 (1994).
2. H. Y. Aboul-Enein, I. W. Wainer, The Impact of Stereochemistry on Drug Development and Use, John Wiley and Sons, New York, 1997.
3. J. E.Rekoske, *AIChE J.*, 47(2001), 2.
4. E. Francotte, P. M. Richert, M. Mazzotti, M. Morbidelli, *J. Chromatogr. A.*, 796(1998), 239.
5. S.L. Pais, M. J. Loureiro, A. E. Rodrigues, *Chem. Eng. Sci.*, 52(1997), 245.
6. M. Pedeferri, G. Zenoni, M. Mazzotti, M. Morbidelli, *Chem. Eng. Sci.*, 54(1999), 3735.
7. M. Schulte, J. Strube, *J. Chromatogr. A*, 906(2001), 399.
8. M. Mazzotti, G. Storti, M. Morbidelli, *J. Chromatogr. A.*, 769(1997), 3.
9. G. Storti, M. Mazzotti, M. Morbidelli, S. Carra, *AIChE J.*, 39 (1993), 471.
10. D.M. Ruthven, C. B. Ching, *Chem. Eng. Sci.*, 44(1989), 1011.
11. F.Charton, R. M. Nicoud, *J. Chromatogr. A*, 702 (1995), 97
12. B.G. Lim, C B. Ching, R.B.H. Tan, S.C. Ng, *Chem. Eng. Sci.*, 50(1995), 2289
13. H. Lorenz, P. Sheehan, A. Seidel-Morgenstern, *J. Chromatogr. A*, 908 (2001), 201
14. M. Amanullah, S. Abel, M. Mazzotti, *Adsorption*, 11(2005) 893
15. G. Strohlein, M. Schulte, J. Strube, *Sep. Sci. Tech.*, 38(2003) 3353.

16. D. M. Ruthven, Principle of Adsorption and Adsorption Processes, John Wiley and Sons, New York, 1984.
17. X. Wang, C. B. Ching, *Ind. & Eng. Chem. Res.,* 42 (2003) , 6171.
18. S.C. Jen., N. G. Pinto, *J. Chromatogr. A*, 662 (1994), 396.
19. F.G. Helfferich, *J. Chromatogr. A*, **768**(1997), 169.
20. M. Mazzotti, G. Storti, M. Morbidelli, *AICHE J.*, 40(1994), 1825.
21. M. Mazzotti, G. Storti, M. Morbidelli, *AICHE J.*, 42(1996), 2784.
22. G. Storti, M. Masi, S. Carra, *Chem. Eng. Sci.*, 44, 1329 (1989).
23. G. Storti, M. Mazzotti, M. Morbidelli, R. Zaciocchi, *Ind. Eng.Chem. Res.*, 34(1995), 288.
24. C. B. Ching, B. G. Lim, *J. Chromatogr.*, **634** (1993), 215.
25. X. Wang, X. J. Wang, C. B. Ching, *Chirality*, **14** (2002), 318.